PRAISE FOR

Miseducation

"Exceptional reporting undergirds the truly shocking facts in this book: the fossil fuel industry is doing all that it can to undermine education about climate change, which will be the most important fact in the lifetimes of kids in school today. Thank heaven for the teachers who stand up for the truth—and thank heaven that this book will spark a crucial national conversation about the hijacking of our educational system."

—BILL MCKIBBEN,
founder of 350.org and Schumann Distinguished Scholar, Middlebury College

"Boy, do we need this book now. As the looming climate catastrophe introduces itself by fire and flood, as the world's leaders need a sense of public urgency to make some hard choices, Katie Worth discovers widespread climate denialism in our nation's schools. Ignorance of the scientific consensus, ideological pressure, fossil-fuel industry disinformation, and a well-meaning but misguided desire to tell 'both sides'—it is a disheartening story, richly reported, clearly told and (we can only hope) just in time."

—BILL KELLER,
former executive editor of the New York Times *and founding editor of* The Marshall Project

"Katie Worth's Miseducation explores an under-appreciated but extremely important aspect of our climate crisis: the active miseducation around climate change in American schools. She explains how conservative politicians, well-funded right-wing foundations, and frightened textbook publishers, have watered down, eliminated, or confused the ways the issue is presented to tens of millions of school children. They hope to raise another generation that will fail to act on what may be the greatest threat to our future. But, as Worth shows, efforts by committed educators has led to some real progress and represents reasons for hope."

—ALEXANDER STILLE,
San Paolo Professor of International Journalism at Columbia University, author of The Force of Things: A Marriage in War and Peace

"In her meticulously researched and vividly written book, Katie Worth provides a detailed, comprehensive, and often enraging examination of the forces that obstruct climate change education in the United States through denial, doubt, and delay. But she also offers a glimmer of hope. *Miseducation* is essential reading for anybody who cares about the climate."

—GLENN BRANCH,
deputy director, National Center for Science Education

"Climate change is an unprecedented threat to our global community, and the frontlines of our efforts to address that threat are in the nation's classrooms where clearheaded, well-informed educators can provide the coming generation with the facts about its causes and likely consequences. But what if those classrooms have been infiltrated by bad actors? In this engagingly written and important book, Katie Worth reveals how the science education that might save us has been influenced by partisan politics and special interests putting the future of us all at risk."

—JOHN L. RUDOLPH,
University of Wisconsin-Madison, author of
How We Teach Science: What's Changed, and Why It Matters

"Young people horrified about climate change are standing up against fossil fuel companies and governments the world over. Amid this global youth uprising, Katie Worth reveals in horrifying detail the ways in which children in American schools are being methodically—and oftentimes successfully—targeted with climate misinformation designed to keep profits and pollution from oil, coal and gas flowing. This deeply reported book names names and reveals filthy secrets and should be essential reading for anybody concerned for the future of humanity."

—JOHN UPTON,
editor, Climate Central

COLUMBIA GLOBAL REPORTS
NEW YORK

Miseducation

How Climate Change Is Taught in America

Katie Worth

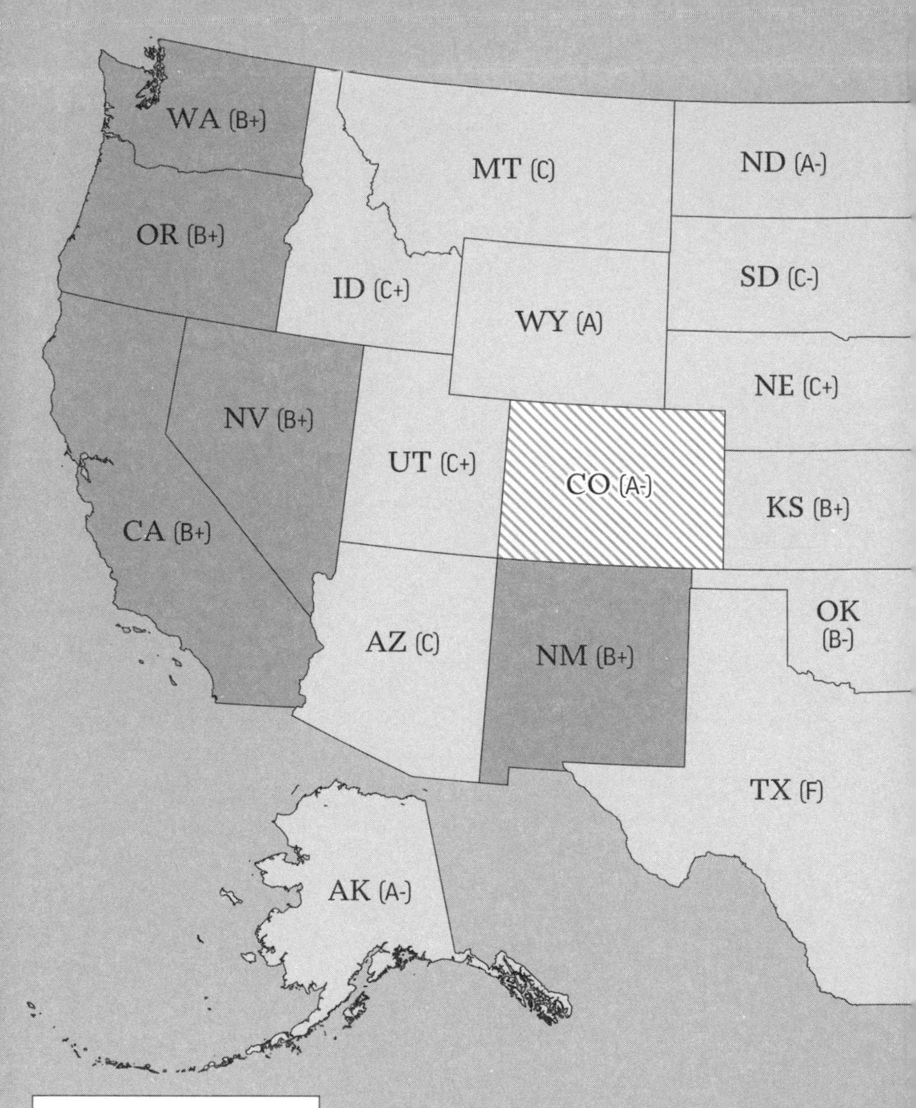

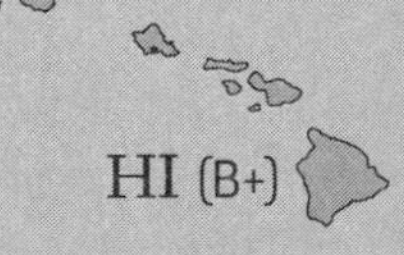

	Dem.	Rep.	Split
A	1	3	1
B	15	6	5
C	0	10	0
D	0	4	0
F	0	5	1

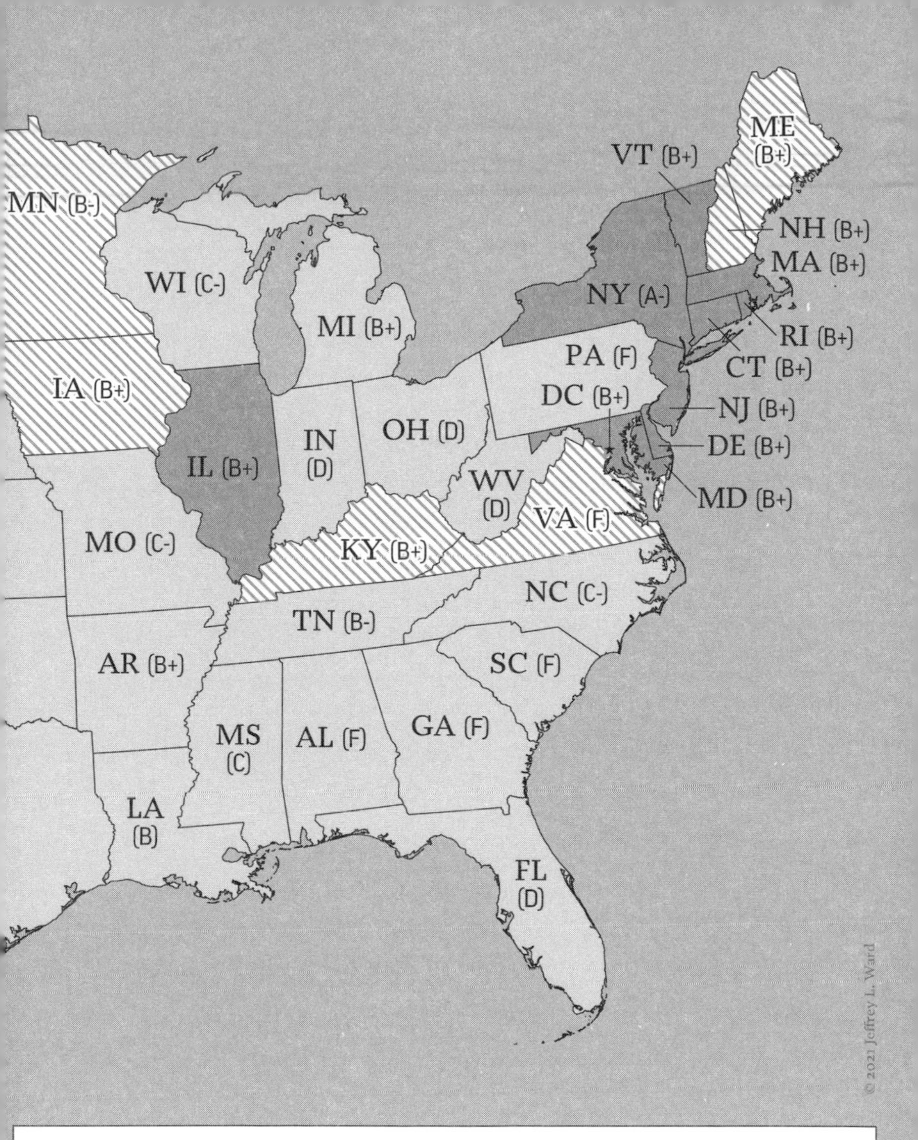

States with a Republican-controlled legislature

States with a Democratic-controlled legislature

States with split control over their legislature
* based on which party won the majority of state elections between 2012–2019

(A)(B) Grades—based on how well the state's academic standards address climate change
* Source: climategrades.org

This book was written with the support of FRONTLINE *and The GroundTruth Project.*

Miseducation
How Climate Change Is Taught in America

Published by Columbia Global Reports
91 Claremont Avenue, Suite 515
New York, NY 10027
globalreports.columbia.edu
facebook.com/columbiaglobalreports
@columbiaGR

Library of Congress Cataloging-in-Publication Data
Names: Worth, Katie, author.
Title: Miseducation / Katie Worth.
Description: New York, NY : Columbia Global Reports, [2021] | Includes bibliographical references. |
Identifiers: LCCN 2021017644 (print) | LCCN 2021017645 (ebook) | ISBN 9781735913643 (paperback) | ISBN 9781735913650 (ebook)
Subjects: LCSH: Climatic changes--Study and teaching. | Environmental education. | Common fallacies.
Classification: LCC QC903 .W683 2021 (print) | LCC QC903 (ebook) | DDC 363.738/74071--dc23
LC record available at https://lccn.loc.gov/2021017644
LC ebook record available at https://lccn.loc.gov/2021017645

Book design by Strick&Williams
Map design by Jeffrey L. Ward
Author photograph by Emma Varsanyi

Printed in the United States of America

CONTENTS

Introduction

Sixth-grade science teacher Kristen Del Real had invited me to come by during her prep period, so for the first time since age thirteen, I found myself walking the halls of my alma mater, Chico Junior High School. The corridors were missing their old rows of lockers—an accommodation for the school-shooting era, I supposed—but still smelled of erasers and turmoil. It was spring 2019, and I had been investigating what American kids learn about climate change, so when I traveled to my Northern California hometown for a visit, I reached out to local teachers to ask how they approach the subject.

I found Ms. Del Real in the 400 wing, preparing a lesson about geological time. She reminded me of many people I know from my hometown: She wore a fleece vest over her jeans, her makeup-free face tan from jogging the oak-lined trails of our local park. She also reminded me of teachers I'd met all over the country. Her demeanor was gentle but authoritative, laugh lines supporting the enthusiasm of her smile.

It was March, and her class wouldn't learn about global warming until May, but unbeknownst to her students, they were already preparing to grasp it. Her lesson on geological time would set them up to understand Earth's natural cycles. Then she would bring in legume sprouts to demonstrate how rhizomes pull nitrogen from the air and turn it into soil nutrients. That would lead to lessons on the atmosphere, solar radiation, the greenhouse effect, and weather systems. "Once all those pieces are in place, when we get to global warming, the kids will just *get it*," she said.

After that would come the part of the year Ms. Del Real loves best: solution projects. For the month of May, her students would work in groups, inventing ways to solve the planet's greenhouse gas problem. "Children are so perceptive. They understand things aren't necessarily great, and it frightens them," she said. The solution projects help dispel that fear, reminding them that "humans are amazing at innovation and invention when we have to be, and the time for that is now."

Three years earlier, her students had been in the middle of their solution projects when they started showing up crabby. Usually, she said, they were excited to get to work. Now, they thought their projects were dumb. "Why are we even doing this?" they asked her. "We don't need to worry about climate change." She soon learned the source of their discontent: Her students had been leaving her lab and walking into history class, where the teacher was showing them YouTube videos alleging that global warming was a hoax, that it was caused by natural solar cycles, and that it was nothing to worry about.

The next day, she walked to the 300 wing and confronted the history teacher about undermining her curriculum. She explained her lessons and methodology, the evidence she has her students examine and analyze, and the California science standards the curriculum fits into. "I said, 'They're eleven. We need to be really mindful of when one adult they trust says one thing and another adult they trust says, "Don't worry about it."' He said, 'Well, I just want them to know both sides.'"

If today is a school day in America, approximately 3 million teachers are educating 50 million children enrolled in 100,000 public schools right now. The scene in each class is playing out differently, since there is no national curriculum. States provide guidelines of what students should learn each year, but schools can use any method they'd like to get them there. Which is to say, it's impossible to definitively describe what kids are learning about recent climate change, since that happens behind the closed doors or on the individual Zoom screens of classrooms in every community in America.

But there's a lot that can be known. To that end, I reviewed scores of textbooks, built a fifty-state database, and traveled to more than a dozen communities to talk to kids about what they have learned about the phenomenon that will shape their future. What I found were points of friction in abundance: Teachers who disagree over whether to teach it. Students who want to learn about it but are not taught. Others who are taught it but reject what they learn. District officials who struggle with teachers who refuse to teach it, or with those who insist on teaching it. Parents who rage that their children are taught it, or that they are not.

That the classroom is not an ideologically neutral space when it comes to climate science is, in a way, strange, because climate science itself *is* ideologically neutral. The evidence for human-caused climate change is now as strong as the evidence linking cigarettes and cancer. Yet—as in the case of the children shuttling between the 300 and 400 wings of Chico Junior High—students are often asked to debate a subject that scientists themselves do not. Adult politics soak into the spongy minds of schoolchildren in a number of ways. Many of the nation's most popular textbooks introduce them to alternate theories for which there is no evidence. Teachers, usually unwittingly, find their way to online lesson plans created by moneyed interests. Some states require a robust climate science education, while others carefully omit it from their academic standards. Every year, lawmakers propose legislation aimed at swaying what children learn about the subject. And, of course, kids hear it outside school, too: One of America's two major political parties—the one that, until recently, held power in all three branches of the US government and still dominates most statehouses—approaches any mention of the climate crisis with something ranging between hesitation and outright denial. Children absorb these messages from the adults in their lives.

It all adds up. Young people are more likely than their parents or grandparents to accept that humans are messing with the climate, but nonetheless, a 2021 UN survey found that a quarter of Americans under eighteen declined to call it an "emergency"—a rate higher than any other nation surveyed in Western Europe or North America.

Why does this matter? Because just as it behooves us to teach students to read or add sums together, we will all benefit if the next generation has basic literacy in the metamorphosing world they find themselves in. Heat-trapping pollution has already begun roiling Earth's natural systems. Among other things, it has unleashed natural disasters with greater frequency and fury than at any time in human memory. Virtually no matter where they live, today's children will bear witness to human-caused climate catastrophes in their communities.

That's certainly the case for Ms. Del Real's students, who in their short lives have already experienced more megafires—fires that burn more than 100,000 acres—than their parents and grandparents ever did. Her classes each year now include a handful of students who once lived in Paradise, a town in the Sierra foothills fifteen miles east of Chico. In 2018, a megafire called the Camp Fire burned 90 percent of buildings in Paradise, earning it the distinction of being the most destructive fire in California history. Scientists generally avoid blaming any individual disaster on climate change, but this one, they say, was covered with its fingerprints. The changes to Earth's atmosphere have shortened California's rainy season both in the spring and the fall; when the Camp Fire caught on November 8, Paradise had received just 0.88 inches of rain in the previous six months. Moreover, California's summers have steadily warmed; Paradise's five hottest summers had all occurred in the five summers before the Camp Fire. That relentless heat had sucked moisture from the town's clay soil and ponderosa pine cover.

As bad as California's fires are today, worse await. If emissions aren't sharply curtailed, a state-funded study found, extreme wildfires will strike 50 percent more often and burn 77 percent more land by 2100. As these fires burn, coastal areas worldwide—places now home to 200 million people—could fall permanently below the high tide line. To survive, those people will have to move somewhere, along with hundreds of millions of others displaced by droughts, storms, and floods. Today's children are likely to watch as catastrophes, displacements, and extinctions tick up with metronomic regularity, transforming their lives regardless of what they once learned in science class.

If preparing children for their own future wasn't reason enough to teach them accurate climate science, these children will soon be decision-making adults, and we know education can powerfully sway those decisions. A study led by climate education researcher Eugene Cordero of San Jose State University followed students who had taken an intensive college course on climate change, and found they made more environmentally friendly decisions than their peers for years after. These decisions—what car to buy, what foods to eat, how to dispose of waste—added up to 2.86 tons less carbon emissions per student per year. Were students across the nation to take a class like this, the paper concluded, the potential reduction in heat-trapping pollution would be about as much as other major mitigation strategies, like rooftop solar and electric vehicles.

This education can be infectious. As science educators Kim Kastens and Margaret Turrin wrote in a 2008 treatise on the subject: "What sets the agenda for public discourse in America—the topics people talk about at the dinner table, the bus stop, the haircutter? The media and popular culture certainly play a key role. But the conversations of today and tomorrow also will be influenced by the ideas and questions that children bring home from school." Sure enough, a study of middle-school children in North Carolina found that parents grew more concerned about the climate crisis after their children were taught about it. This intergenerational effect played out most strongly in conservative families. Daughters were especially influential, and fathers were especially influenced. Furthermore, young people who do understand the climate crisis have proven capable of leading the rest of us in action. In 2019, more than a million students worldwide skipped school to bring adult eyes to the crisis. Teen activist Greta Thunberg is now a towering figure in the movement to slow emissions. In Eugene, twenty-one young people aged eleven to twenty-three sued the federal government for not protecting their future.

Classrooms have emerged as a battleground in the American political war over climate change because what kids learn about climate change now will directly impact the speed and ambition of action taken for decades to come. That in turn will decide the quantity of fossil fuels extracted from Earth. If a significant portion of young people grow up to doubt the climate crisis is real, as their elders do, little is likely to change. The inertia of the status quo is so high that even a modest dose of doubt inoculates against action. After all, who can justify removing a pillar of our economy without certainty that it is necessary, and that it will work? Confusion and doubt over the causes and impacts of global warming could reign in American politics another three years or another thirty years. That difference is a matter of trillions of dollars for the fossil fuel industry, and of accelerating chaos for the planet.

How did we get here? Why are millions of American children learning mixed or false messages about the phenomenon that will dictate their future? How did there come to be a red–blue divide in climate education? Who has tried to influence what children learn, and how successful have they been? I spent years tracking down the answers to these questions. What I found were the unmistakable signs of moneyed interests and entrenched ideology. Fossil fuel lobbyists, flaccid textbook companies, networks of free-marketeers and evangelical leaders, and the American political machine have each had a role in the widespread, calamitous, and in some cases, intentional miseducation of our children. It's safe to say that across the country, intrepid teachers rigorously educate their students about climate science. It's also safe to say that, commonly, a teacher down the hall is miseducating them about it.

Ms. Del Real prevailed over the history teacher. After their confrontation, he begrudgingly agreed to stop showing the climate hoax videos to her students. But it is unlikely to be the last time those children run into an adult with his views. Ms. Del Real teaches in what she calls "a mixed community politically." Chico has the crunchy liberal trappings one would expect of a California college town. But it sits in a deeply conservative agricultural county. Her students often arrive with preconceived and unsupported notions about climate change. She is careful to treat those notions with respect, but also to submit them to the rigors of scientific thinking, as she would any other subject. Usually, by the time they get to the solutions projects, even the doubters understand the implications of a changing atmosphere and are eager to dream up answers. She's sure that one of these years they'll come up with something life-altering. "I really believe that eleven- and twelve-year-olds can save our world," she said, a broad smile underscoring her optimism. "They're brilliant."

The Science and the Doubt

In Springdale, Arkansas, students interested in learning about recent climate change have a few things going for them: Their state science standards require that they learn about it. Their school uses modern classroom materials. And one in eight children in their district are Pacific Islanders, whose families hail from a part of the world where the impacts of climate change are well defined and world famous.

But they could face the problem of Mr. Nokes.

For the last decade, Charles L. Nokes has taught five classes a day of environmental science to tenth, eleventh, and twelfth graders at Har-Ber High School. When I met him, he wore a knit sweater a few shades darker than his pearl-silver beard; his bespectacled blue eyes lit affectionately each time a student meandered into his class to chat during their lunch break. As a fourth-generation farmer, Mr. Nokes certainly cares about the environment. In fact, he once won an award from the Arkansas Cattlemen's Association for preventing his Ozarks ranch from polluting a nearby river. But he considers environmentalism too radical, preferring the idea of conservationism. "A conservationist believes that we should be good stewards of what we have to work with," he said. He grew up on his family's ranch, and then served in the military. After, he started his own consultancy and traveled around the world advising governments and businesses on how to resolve environmental and food-production problems. He might have spent his whole career on that work were it not for a serious personal loss. His wife passed away young, and her death left him a single parent. "I had two children who needed me and I discovered I was way too busy," he said. "I elected to change my life rather dramatically, and this"—he gestured at the classroom we sat in—"is the result. I don't regret it for a minute."

When Mr. Nokes first learned about climate change, he found the idea very concerning. But he also doesn't like to blindly trust what scientists tell him. He gave me a personal example, based on a time when he was involved in the egg industry. Readily available and full of protein, eggs had long been considered "the most perfect food that humans knew," Mr. Nokes said. Then, in 1968, the American Heart Association recommended that people eat no more than three eggs per week, since egg yolks are loaded with cholesterol and when human hearts clog with cholesterol they are prone to failing. Early research seemed to indicate that cholesterol in eggs might indeed lead to cholesterol in the heart. Doctors, government officials, and health gurus bought the theory, and soon it became conventional wisdom. But it never squared for Mr. Nokes, who had seen his ninety-year-old grandfather eat two eggs every morning his whole life without apparent consequence. It took decades before countervailing research convincingly showed the connection between dietary cholesterol and human cholesterol to be far more complicated than first imagined, and for health agencies to stop definitively linking eggs to cardiovascular disease. “For that period of time, they destroyed the egg industry,” Mr. Nokes said gravely.

Mr. Nokes does not take scientists’ conclusions on climate change as truth, because he worries the same oversimplification and bandwagon-jumping that beset the reputation of eggs might also afflict climate science. And, he told me, flaws in the science might be more difficult to detect, because that science involves predicting the future of an exceedingly complicated system using models. It’s not that he doesn’t trust anything climate scientists say. He concedes that carbon dioxide levels are increasing. “But what impact does that have on global climate? I don’t have a clue,” he said. The recent rise in global temperatures may be caused by solar cycles or gravitational forces instead of humans, he told me. “Like so many things, including an egg, there’s just more to be learned before we know for sure what we’re even measuring,” he said.

What Mr. Nokes did not acknowledge was that there’s a fundamental difference between the trajectories of egg research and climate research. The more that scientists began studying the egg-heart link, the fuzzier and more nuanced their conclusions became. Scientists are still untangling the precise connection between the cholesterol that humans eat and that which shows up in our bloodstream. Not so with climate. The more that scientists have studied a link between human industry and global temperatures, the more unambiguous they have found it. A study examining 12,000 peer-reviewed papers (papers scrutinized and vouched for by other scientists) published between

1991 and 2011 found 97 percent of those that expressed a position on the cause of climate change agreed that it was happening because of human activity. A follow-up study looked at the 3 percent that disagreed, and found those studies were riddled with methodological flaws. Today, looking for a peer-reviewed study suggesting that humans are not warming the climate is like searching for an earthworm in a henhouse. A review of global warming papers published in the first seven months of 2019 found zero.

Why such certainty? Mr. Nokes is correct that the planet's climate system is so complex that we don't yet fully fathom it. But the mechanism that has driven the climate to warm is elegantly simple. Certain gases in the atmosphere reflect heat back down to the planet, effectively trapping it on Earth. Modern industry, mainly through burning fossil fuels, has increased the concentration of one of those gases, carbon dioxide, by nearly 50 percent. We don't need to understand the connection between chicken eggs and cardiovascular health to conclude that eating thirty-four in a sitting will result in a bellyache.

Nor is the science newfangled. The idea that some ingredients of the atmosphere keep heat from drifting into outer space was first floated in the 1820s. Forty years later, an Irish born physicist theorized that changing the proportion of those ingredients would change Earth's average temperature. The degree to which that was true was established in 1896, when Swedish chemist Svante Arrhenius undertook thousands of "tedious calculations" and concluded that halving the concentration of carbon dioxide in the atmosphere would drop Earth's mean temperature by several degrees Celsius, while doubling it would do the reverse. In a lecture that year, he offered the good news that at the rate humans were burning coal, we would increase Earth's temperature significantly over the next 3,000 years. He invited the audience to "indulge in the pleasant belief that our descendants, albeit after many generations, might live under a milder sky and in less barren surroundings than is our lot." In 1938, an amateur climatologist named Guy Callendar gathered weather data and, through his own tedious calculations, realized Arrhenius had his theory right but his timeline quite wrong: In fact, modern industry had already begun to warm Earth. Like Arrhenius, he saw this as "likely to prove beneficial to mankind."

Over the coming decades, however, scientists began to doubt whether turning up Earth's thermostat was an especially good idea. Clues from the geologic record indicated that high levels of atmospheric carbon correlated with sea levels that would drown much of the populated world. By 1965, the scientific community found the warming climate worrisome enough to produce a report and brief President Lyndon B. Johnson, stating the burning of fossil fuels "may be sufficient to produce measurable and perhaps marked changes in climate" by the year 2000. Johnson, in turn, was alarmed enough that in a special message to Congress, he noted that "this generation has altered the composition of the atmosphere on a global scale through . . . a steady increase in carbon dioxide from fossil fuels." Scientists started slicing ice from ancient glaciers and measuring the prehistoric carbon dioxide trapped in them. In 1987, for instance, a team drilled a cylinder out of a 160,000-year-old Antarctic ice sheet and found the carbon dioxide levels had drifted between about 190 and 280 parts per million for that entire period. The only deviation from that range had occurred in just the last few decades. The summer before, the modern atmosphere's carbon dioxide concentration had measured more than 350 ppm for the first time. Today it is well over 400 ppm.

The implications were stunning and frightful, but would not occur for several decades, raising another problem. In their seminal book on the spread of climate denialism, *Merchants of Doubt*, Naomi Oreskes and Erik M. Conway interviewed one researcher who spent the late 1970s trying to get Washington policymakers to pay attention to the crisis. But when he told them that carbon dioxide would have major impacts on the planet in fifty years, they answered: "Come back in forty-nine years." Of course, in forty-nine years, the carbon would already be in the air, an egg that could not be uncracked.

In an attempt to take action before the forty-ninth year, the United Nations convened the world's top climatologists to consider the problem. In 1990, the resulting Intergovernmental Panel on Climate Change issued its first report. There were still a lot of details to be sorted, but the panel said it was certain of the following: The natural greenhouse effect keeps Earth warmer than it would otherwise be, and human activities are magnifying that effect. In the years since, the group has come to include hundreds of climate scientists, who have collectively generated reports that are ever more specific and dire in their statements. Recently, the reports include not just predictions, but also descriptions of changes that have already transpired.

To name a few such effects: The world has already warmed by about a degree Celsius, and the five hottest years in human history have all occurred in the last five years. Those years would have been even warmer were the oceans not absorbing some of the heat, so it is no surprise that the oceans' five hottest years have also occurred in the last five years. The ice sheets covering Antarctica and Greenland have both lost ground, especially since 2009. Sea levels have risen eight inches since 1900, and the speed of that rise is curving upward like the profile of an eggshell.

Presented in a list of statistics, the changes can seem abstract, but American climatologists have warned they are not confined to a faraway land or a distant future. California's summertime forest fires have increased in size more than fivefold over the last fifty years. In New Jersey, the sea has been rising about one inch every six years, forcing some neighborhoods to retreat from their beaches. Colorado ski towns are seeing their busy seasons get shorter. The warming Great Lakes have hosted toxic algae blooms, jeopardizing the drinking water of millions. In Arizona, heat-related deaths have been on the rise. And even in Arkansas, research shows, the soil around Mr. Nokes's farm has begun to dry. Rainfall now often comes in heavy downpours, causing floods. All of this will get worse.

If the connection between fossil fuel pollution and climate change were truly unclear, I asked Mr. Nokes, why would oil and gas companies now all admit that it's real? "They're very astute businesspeople, that's what I think," he said. "They have their ear to the ground, so to speak." On this we could agree. Thanks to a wealth of investigative reporting in recent years by *Inside Climate News, the Los Angeles Times*, Columbia Journalism School's Energy and Environmental Reporting Project, and the *New York Times Magazine*, as well as the historical research in Oreskes and Conway's book, we know a great deal about the industry's astuteness. In 1957, researchers at Exxon-predecessor Humble Oil published a study discussing the fate of the industry's carbon emissions. The 1965 Johnson briefing included research by a scientist at Shell. In 1968, the American Petroleum Institute funded a report about air pollution, which discussed the mechanisms and risks of global warming at length, concluding that "there seems to be no doubt that the potential damage to our environment could be severe."

By the late 1970s, eager to understand more about the phenomenon and its implications, Exxon began hiring scientists, collaborating with academics, publishing peer-reviewed studies, and running models on supercomputers. At the time, their conclusions were in step with mainstream scientists. "Present thinking," said one Exxon scientist in a 1978 briefing to company management, "holds that man has a time window of five to ten years before the need for hard decisions regarding changes in energy strategies might become critical." Exxon even outfitted one of its largest supertankers with custom-made instruments, hoping to pinpoint how much atmospheric carbon the oceans were absorbing.

Other fossil fuel companies were making similar moves, and in 1979, the American Petroleum Institute, the nation's largest fossil fuel industry group, established the "CO2 and Climate Task Force" to facilitate collaboration. Members included Amoco, Chevron predecessors Standard Oil of California and Gulf Oil, Exxon, Mobil, Phillips, Shell, Sunoco, and Texaco. At the time, the industry was open to taking responsibility for the problem. According to the minutes of one meeting held in February 1980, a Texaco representative suggested the Task Force should consider the "overall goal" as helping to develop ground rules for carbon emissions and their cleanup. For a period, industry researchers became such highly reputed experts on human-caused climate change that the scientific community sought their contributions to reports and their presence on panels.

But as models began to show more dire outcomes, the industry became less forthcoming about its research. In the early 1980s, the global warming primers passed around to executives began to be marked, "Not to be distributed externally." Even as the industry was spending millions of dollars on climate research, it excluded the conclusions of that research from federal filings or reports to security regulators. In the mid-1980s, Exxon abruptly defunded the bulk of its research, laying off many of its climate scientists and halting publication on the subject in peer-reviewed journals. By then, the API had also disbanded its taskforce. The industry seems to have decided that what it needed more than scientists was lobbyists.

That's because politicians were finally starting to seriously weigh the idea of regulating carbon emissions. In 1988, discussions of an international treaty began. That summer, in the midst of a deadly heat wave, a NASA scientist testified before Congress that global warming had already begun—a quote that showed up on the front pages of the *New York Times* and many other newspapers. The industry feared that radical change could be coming and that the financial stakes would be tremendous. Trillions of dollars worth of fossil fuels remained in the ground, and regulation could leave them there. After the NASA scientist's testimony, an Exxon public-relations officer wrote in a memo that it would be best for the company not to highlight the connection between fossil fuels and climate change. "Emphasize the uncertainty in scientific conclusions," he wrote.

It was not a strategy that Exxon officials developed themselves. They adopted the same tactics that the tobacco industry had used to hide their products' link to cancer. The idea was to raise questions about the validity of the science, which would cause the public to doubt whether government regulation was warranted. Not everyone had to doubt—just enough people to stymie action. As Oreskes and Conway lay out in *Merchants of Doubt*, this strategy works because while policymakers hesitate to take sweeping action without moral certainty, scientists have an allergy to definitives. Scientists describe even ideas supported by overwhelmingly persuasive evidence as "theories," and perpetually test those theories for weaknesses. But just because the work of science is never done doesn't mean its products can't be trusted. We may not know everything about plate tectonics, but neither do we throw up our hands in bewilderment every time the ground shakes. The scientific process brings a picture into focus. When something has been studied as much as climate change, the remaining work isn't determining whether the picture is a landscape or a still-life, but rather finding and sharpening the fuzzy lines. Meanwhile, climate deniers have worked to magnify those fuzzy lines in an attempt to make the entire picture appear blurry.*

* A note on terminology: This book uses the term "climate denier" to describe groups or people who know (or who should know, based on their position or declared authority) that the conclusions of modern climate science are legitimate, but who nonetheless promote the idea that they are not. For members of the public who have been misled by the sources of information they trust, the term "climate doubter" applies. Also, this book uses the terms "climate change," "climate crisis," and "global warming" as shorthand for "anthropogenic climate change," the extraordinary transformations in our atmosphere and ecosystems that have occurred since industrialization. Any reference to natural climate change is described as such.

In 1989, ExxonMobil, Shell, British Petroleum, and other leading fossil fuel organizations formed a lobbying partnership that, in an inspired stroke of doublespeak, they named the Global Climate Coalition. The group and its affiliates would spend the coming decade launching public attacks on the integrity of leading climate scientists, spending millions on advertising to fight action on the issue, and funding an array of front groups that promoted doubt. Their mouthpieces cherry-picked data and exploited outliers in the research to exaggerate uncertainty that, to scientists, was diminishing by the day. Less than a decade after Exxon's scientists were publishing peer-reviewed studies showing that humans were drastically altering the atmosphere, Exxon's board stated publicly that the science was too murky to warrant action. Industry efforts ramped up further as the UN negotiated the 1997 Kyoto Protocol, which aimed to curb carbon emissions that would have vast economic consequences for the industry. Kyoto's hopes of passing in the US were already dim, as Democrat and Republican politicians were united in their hostility toward it. As it was being negotiated, the US Senate voted 95–0 on a resolution opposing any treaty that asked less of developing countries than it did of the world's richest countries, a key provision of the proposed protocol.

A few weeks after Kyoto was finalized, the API hosted a meeting attended by representatives of Exxon, Chevron, utility giant Southern Company, and a handful of conservative think tanks. They met to strategize around an explicit and singular goal: "Victory will be achieved when average citizens 'understand' (recognize) uncertainties in climate change," declared the memo that came out of the meeting.

The memo contended that the Clinton administration and environmentalists had "conducted an effective public relations program" to persuade Americans that the threat of global warming was real. Worse, the public had been "highly receptive" to that message. Fortunately, the memo said, a recent survey had suggested that public opinion was not yet fixed about climate change, and with sufficient investment, those open minds could be changed. The group laid out plans to expand the already steady stream of climate-denying materials for distribution to the media and politicians. It would host a series of college debates over whether human-caused climate change was really happening. A sympathetic TV journalist would be persuaded to critique mainstream climate science.

In addition to politicians, media, and industry leaders, the meeting attendees had another group in their sights: children. The memo laid out plans to target schoolteachers and students, so as to "begin to erect a barrier against further efforts to impose Kyoto-like measures in the future." A new "Science Education Task Group" would reach out to the National Science Teachers Association and other influential organizations to help develop school materials and distribute them "for use in classrooms nationwide." The matter was urgent: "The teachers/students outreach program will be developed and launched in early 1999."

But then something went sideways. The memo was leaked to an environmental organization and then to a reporter at the *New York Times*, who wrote a front-page story about the group's intentions. Scientists and activists blasted the nascent campaign as a cynical attempt by monied interests to manipulate the public. The response was strong enough that the people who had attended the meeting told the reporter the plans were only "very, very tentative." The group, seemingly, went quiet.

That did not mark the end of industry efforts, though. "Let's face it: The science of climate change is too uncertain to mandate a plan of action," read one of scores of ads the oil industry placed in newspapers. Exxon's chief executive, Lee Raymond, seemed to make it a personal mission to spread this message: In a talk in Beijing, he falsely claimed that the earth had cooled over the previous twenty years, and argued, "Let's agree there's a lot we really don't know about how climate will change in the twenty-first century and beyond." In an annual meeting, he said "projections are based on completely unproven climate models, or, more often, on sheer speculation." After President George W. Bush and Vice President Dick Cheney took office in January 2001, Exxon's top lobbyist pushed for Clinton-hired scientists "with aggressive agendas" to be fired and replaced with hand-picked industry-friendly officials, and the new administration complied. Later that year, Bush killed any remaining hope of ratifying the Kyoto Protocol.

The industry's doubt campaign seemed to extend everywhere—except its own ventures. While it was shining light on supposed uncertainties in climate science, it quietly incorporated that very science into its business models. Companies began to build oil platforms taller to withstand predicted sea level rise, and stronger to weather bigger storms. Some assessed the risk that thawing permafrost posed to their buildings and pipelines. Exxon explored the possibility of making new profits as Arctic ice melted and exposed long-buried oil stores.

The industry had successfully stymied action on climate change, but not without a hit to its reputation. Public and shareholder pressure mounted. Environmentalists organized protests and boycotts. In 2006, the UK's science academy sent a harsh letter to Exxon accusing it of being "inaccurate and misleading." Finally, in 2008, Exxon said it would stop funding groups that dispute climate science.

By then, however, the doubt campaign had proved phenomenally successful. Long after the Global Climate Coalition dissolved and fossil fuel companies quieted their overt doubt campaigns, ramifications persist. President Donald Trump was applauded by his party when he withdrew the US from the latest attempt at a UN climate accord. Deniers riddled his appointments to federal agencies. A fair portion of Americans believe there are scientific "sides" to the issue, when science isn't about "sides" at all—it's about evidence. Dismissal of climate change has become so entrenched in conservative ideology and identity that the greatest predictor of whether someone trusts the conclusions of climate science is not their science literacy, but their political affiliation. A 2020 survey found the partisan divide over climate change was greater than any other modern issue it asked about—greater than immigration, gun policy, race, or the economy. As American adults have sopped up the industry's message about climate change, they have squeezed it into the minds of children.

Mr. Nokes patiently enumerated to me a dizzying list of reasons he questions modern climate science. I had heard them before, mostly at conferences hosted by climate denial groups. I had also heard answers to those questions from scientists I had met through my reporting. Mr. Nokes dismissed the models as too simple for a complex system. (In fact, global models have so far proven cunningly accurate. A 2019 study assessed seventeen major forecasts from 1970 to 2001 and found ten of them precisely predicted where we are today. Five others had faulty assumptions about how much pollution humans would emit, but once those were corrected, they, too, were accurate.) He suggested that the recent warming could be part of a natural cycle. (The models take natural cycles into consideration and still conclude humans are to blame for the warming.) Or, he said, it could be caused by solar flares. (NASA data shows that solar output has not significantly changed in decades.) He questioned the quality of historic temperature records. (Climate historians have gathered a variety of data sources to control for variations in quality.) He said that plants would thrive with higher concentrations of carbon dioxide. (This is true but does not solve the multitude of other problems.) And finally he argued that Mother Nature will eventually rectify herself, no matter what we do. (Also true, but in the meantime, the suffering will be vast.)

I asked Mr. Nokes if he had read the UN's latest IPCC report. He said he had.

"I don't give it a lot of credence on the surface," he said. "The conclusions seem to me to be a little bit political."

I asked him what he made of the broad scientific consensus on the issue.

"It's not based on real science," he said.

I asked him how he talked about all this with his students.

"Just like I'm talking to you," he said.

I asked him how much longer he hopes to teach.

"Until they drag me out," he said.

The Teachers

Once, on an island little wider than the length of a blue whale and no taller than the wingspan of a giant Pacific octopus lived a nine-year-old named Izerman, whose mind teemed with stories of the creatures he had met.

"Once, I saw a manta ray flying up. It was leaping from the water for five seconds!" he told me and a colleague who were visiting his island, his hands flying through the air to illustrate. "It was one of the most shocking things I've ever seen."

"Once, I went to Arno; I saw a little red thing in a tree," he said a moment later. Arno is an island thirty-five miles from the one Izerman lives on. "I got really scared and I yelled, '*scorpion!!*' My friend was like, 'Where, where, where!?' Then he grabbed a bottle and we catched it."

"Once, when I was swimming, I saw a large eel. It was greenish-orange with teeny tiny black spots. My cousins ran away while I was chasing it." He followed the eel until—plot twist—it led him right up to a shark, and then it became Izerman's turn to run away.

Then came the stories of the smartest bird he ever met, of the parrotfish he always swam out to visit on the reef, and of the time he put a lizard in his mom's bed as a prank. Then he gravely told us of the time he made a misjudgment any young zoologist might make, concluding, "From that day on I've never kissed a crab."

When Izerman was six years old, he was watching a YouTube video about the lives of Arctic seals and he heard the words "global warming." He asked his mother what those words meant. She told him that when too much pollution goes into the air, it creates a kind of blanket that warms up the earth. It was not the last Izerman would hear about global warming. He heard it in school the next year, and then the next year and the next. One day, we watched as Izerman and his classmates sat cross-legged on the floor in a semicircle around their fourth-grade teacher, Waisake Savu. Mr. Savu pointed at colorful handmade posters taped to the blackboard.

"What do you know about the North Pole and the South Pole? What do we have there?"

"Melting ice," chorused Izerman and his classmates.

"Okay, yes," Mr. Savu said. "And as the ice melts, what happens to the—"

"The water level rises," said the class in unison, before he even finished.

Rising water levels are of special interest to Izerman and his classmates because the island they live on—Majuro Atoll, capital of the Marshall Islands—sits at an average of 5.9 feet above sea level. Its highest natural point is 11.5 feet. The tallest pile of trash in the island's landfill is higher. Majuro, like all atolls, is made up of a series of slender islets that chain together into a broken ring encircling a lagoon. What causes atolls to form once mystified geologists. Charles Darwin himself was puzzled by the question on his voyage around the world in the *Beagle*. His best guess was that atolls were once coral reefs surrounding volcanoes; eventually, the volcanoes sank under the sea, while the reefs rose above it. Scientists now agree he was right, though the specifics remain a mystery.

If you were to point at the spot on the globe roughly equidistant from Hawaii, Australia, and Japan, you'd wind up in the neighborhood of the Marshall Islands. The Marshallese origin story depends on who you ask, but most versions involve a god named Lowa who lived, lonely and bored, on a primeval sea. According to one iteration, he began humming to himself, and his humming brought into existence the reef and sandbanks. He hummed some more, creating plants and animals. From his leg emerged a man and a woman. They put the islands Lowa had created into a woven basket and then placed them into two lines in the ocean. The eastern chain of islands is named Ratak, after the sunrise, and the western one Ralik, after the sunset. Including Majuro Atoll, which sits low in the sunrise chain, the Marshall Islands comprise twenty-nine atolls and five islands. Their combined landmass is about that of Washington, DC's, but they are spread across an expanse the size of Mexico. People have lived on these islands for at least 2,000 years.

The islet Izerman lives on is so narrow that when he stands on the road above his house, he can see both the ocean and the lagoon. His house is on the lagoon side, and between it and the water is a cement seawall, meant to protect against floods. Sometimes, though, high tides combine with surges from passing storms to make king tides. "You see those waves there?" Izerman pointed to the lagoon. "Imagine them wayyyyy bigger." At their worst, these king tides can overcome the wall, the beach, the house, and the entire island, salting the groundwater, killing plants, disinterring corpses from graveyards, and knocking down buildings. Izerman's elders remember king tide floods coming once a generation. When we met him, Izerman had witnessed four of them in his young life. Once, he said, he was sitting on the seawall during a king tide. "I didn't see the wave coming. It just pushed me off and I thought it was something else than a wave. I thought it was, like, a monster." By the next morning, the seawall he had been sitting on was bent over. The waves had picked up a neighbor's canoe and used it as a battering ram against a covered patio Izerman's family had just finished building.

Since tide gauges were first installed in Majuro in 1968, the seas there have risen six inches. A portion of this rise is due to the natural, cyclical sloshing of seawater from one end of the Pacific to the other, but the majority has been caused by the swelling and warming oceans—and that part is accelerating. Another foot or so of sea level rise would bring king tides on an annual basis—too frequently for the island's residents to fully recover before the next one hits. The scientists we talked to said the seas could permanently rise this much by the 2050s. In the meantime, they predict that major storms in the Pacific will strike less frequently, but those that do will be more ferocious. Izerman worries that one of those could bring a king tide so monstrous that "the only things that would be sticking up is coconut trees." He has come up with a plan: "If I learn how to climb trees, maybe if big waves come, I can climb in a tree and wait until it gets shallower." On the other hand, he reasoned, he might not get to deploy that skill more than once. "My parents will just take me to America, and there, there won't be any trees to climb, so it's probably not going to be useful."

When we met Izerman in 2017, the family had just spent part of the winter in Enid, Oklahoma, visiting extended family and pondering what it would be like to live there instead of Majuro. The Marshallese have the option to live and work in the US without a visa, as part of a special agreement between the two nations that allows the American military to maintain a base in the islands. Since 1995, more than a third of Marshallese people have migrated away from their nation, most looking for better education, health care, and employment than is available in their home islands. Some are also motivated by the bleak climate predictions for their homeland. When Izerman's family got home from Enid, they took to debating the idea of a move to Oklahoma over family meals. Izerman's parents would rather not leave their home islands, but constantly wonder if their children would have a better future in the US. Izerman's older sister favored the idea, but Izerman argued forcefully against it.

"Once, when I was in America, I was looking for animals. The only thing I found in the whole month was a cricket. I let that cricket go because I knew it was hibernating," he said. "So it's going to be hard, living there."

I traveled to Enid, Oklahoma, to find out what Izerman might learn in the schools there about climate change. Enid sits at a modest crossroads about an hour and a half north of Oklahoma City. Enid is like the Marshall Islands in some ways: It is home to about 50,000 people, it is predominantly Christian, it hosts a US military base, and many of its residents love the land they live on and, through cultivation, know it intimately. There the similarities stop. Whereas in the Marshall Islands, one is never more than a five-minute stroll to the ocean, Enid sits roughly halfway between the Pacific and the Atlantic; strolling would be out of the question. Whereas in the Marshall Islands, the temperature has never been recorded below 71 degrees, Enid has four seasons. Whereas in the Marshall Islands, bounty comes from the sea, Enid draws its bounty from the land. In fact, among wheat aficionados, Enid enjoys a measure of fame, since it has at times boasted the largest grain storage capacity of any town in America. In the sunshine, the wind makes the wheatfields around Enid shimmer and ripple in a way that evokes Majuro's lagoon. In some fields, giant three-armed windmills collect the moving air and turn it into electricity. Sometimes toiling right next to them are pumpjacks, muscular motorized metal arms that pull several gallons of crude oil from below the plain with each pump. But the wheat fields are not as eternal as they seem. Oklahoma has grown considerably drier than it once was. Soon, wheat farmers will have to irrigate their land—something most have never done—but there will be less water available to do so. If they do not irrigate, their crop yields will drop by half.

The Enid origin story depends on who you ask, but one version says a railroad official named it after a heroine in a Tennyson poem whose life is plagued by a jealous husband. Others say that an early Oklahoman restaurateur accidentally hung a "DINE" sign upside down and it stuck. For centuries prior, the region had been considered undesirable for white settlement, so in the 1830s, the US government forced five Indian tribes off their much-more-desirable land in Georgia, Alabama, Tennessee, North Carolina, and Florida, and marched them to Oklahoma on what one tribal leader famously described as a "trail of tears and death." The Treaty of New Echota, signed in 1835, defined the tract around Enid as "the Cherokee Outlet," to be a perpetual territory of the Cherokee nation. Fifty years later, the tract started looking desirable to whites after all, so the US Interior Secretary reread the old treaty and decided it should not have granted the Cherokees the land in perpetuity. He threatened to confiscate the land if the tribe did not sell it back to the government cheaply so that it could be given away to whites. The tribe complied. On a September day in 1893, more than 100,000 land seekers stood on one side of an invisible line, until, right at noon, a volley of shots was fired. Thus began the largest land run in American history. People on horses, buggies, and even bicycles raced to claim a free homestead. Enid was founded that very day.

It wasn't for another eighty years that the Marshallese would arrive. Some of the first came to Enid to study at a small Christian college called Phillips University. As is common in immigrant communities, those first immigrants attracted more. By the time that college closed due to financial difficulties, the Marshallese had formed a community in Enid. Many new arrivals from the islands began working for the AdvancePierre Food Company, known for producing prepackaged school lunch food. The Marshallese fit into Enid, in the sense that they were churchgoing people and Enid is a churchgoing town. Today, there are at least eight Marshallese churches in Enid. The Marshallese community accounts for about a tenth of the student body at Enid High.

The day I visited Enid High, I climbed three floors to the office of Lori Palmer, who is somehow even friendlier than the average teacher. ("I'm *such* a people person," she gushed.) Ms.

Palmer grew up in a town outside of Enid where she attended a high school with a graduating class of thirteen. She had never even heard of the Marshall Islands until children from there began showing up in her reading classes. Now she serves as the high school's English Language Learner coordinator. Her office is covered in woven baskets and shell-laced jewelry, gifts from Marshallese students. She has eight ukuleles in her office and they are deployed on a daily basis, sometimes all at once. For Ms. Palmer, teaching is about creating bonds with kids. "They're not going to learn from you if they don't think you care about them," she told me. To that end, she is in constant text-contact with the 200 English language learners she is charged with. When I showed up, she group-texted the leaders of the school's Islander Club, which she advises. They were delighted to be pulled out of class. "I'm surprised to see you here!" she said to one, whose mom had just given birth to a baby girl the day before. "I walked here from the hospital," the student said, adding sheepishly, "My mom wanted me to come to class because she's worried about my grades." Ms. Palmer rolled her eyes and chided, "She should be!"

I asked the five Marshallese students who had gathered if they had ever learned about climate change in school. Senior Lana recalled hearing about it in eighth-grade history, when the teacher showed the class a short video about it. The other four said I was the first adult to ever talk to them about climate change on school grounds. Seventeen-year-old junior Eve was born and raised in Enid. She hasn't visited her home islands since she was one year old, but speaks vividly about what it is like there. I asked her what she knows about climate change.

"The world's getting hotter. Glaciers are melting. Sea level is rising. Animals are dying. Australia is burning," she answered matter-of-factly.

I asked her how she knew all that, if it hadn't come up in class.

"I researched it myself. Googled it. Because I want to know what's going on in the world," she said, her voice tinged with adolescent scorn.

In defense of the school, Ms. Palmer told us that global warming was not a requirement of Oklahoma's educational standards—the state's expectations of what students should learn in each grade and subject. Those standards, written by a panel of educators, approved by Oklahoma's Board of Education, and reviewed by the state legislature, made no mention of recent climate change in required classes. I asked Eve how she felt about that.

"It's kind of disappointing. Because, like, a real thing is happening," Eve said.

What Ms. Palmer said about the state education standards was true—climate change did not come up in any required class. But some schools offered electives that did cover it. After I left Enid, I drove to the Oklahoma City suburb of Edmond, where I visited a school that offers one of those electives. Veteran science teacher Andrea C. Sampley met me at the front desk of Edmond Memorial High School, and then navigated through the halls so efficiently that I had to speed-walk to keep up with her. She slid to a stop on her flats when she reached the door of her classroom, where she teaches advanced placement environmental science, including a unit on climate change. Every year when the topic arises, a handful of her students "buck up and get argumentative about it," she said. I asked her how she handles a student who thinks what she's teaching is baloney.

"What I've found very effective with students is to start with the historical timeline," she said. "If you look at average temperatures over time, you see it takes hundreds of thousands of years for things to drift up, things to drift down. Then when I show them the temperatures during the itty-bitty, teeny-tiny time period since the beginning of the Industrial Revolution, the students start going, 'Waaaait a second.'"

Ms. Sampley had taken over the class a few years earlier from a teacher who had a radically different approach. "I probably shouldn't talk about that," she demurred. I pressed. "I'll just say that I have heard another AP environmental science teacher say that they absolutely do not believe in climate change. And that they had that stance because their family was in the oil and gas industry." I asked how the students could possibly have prepared for the advanced placement exam—a national test for which students can get college credit if they score well enough—if they couldn't answer its many questions about recent climate change. "The teacher gave a packet about climate change to the students and they worked on it on their own," she said.

How many American students get a robust education about climate change in school? How many learn from a climate-change denier? And how many have an experience like Eve's in Enid, where they learn nothing unless they google it themselves? Eric Plutzer, of Pennsylvania State University, has led a pair of surveys in collaboration with the National Center for Science Education to get at these questions. The most recent, conducted in 2019, surveyed 1,427 science teachers across the country. It found that most public school students hear the words "climate change" uttered in a classroom at some point, since most science teachers bring up the subject during the school year. That's not to say the students are hearing it a lot. About half the science teachers surveyed spent between zero and two hours on recent climate change—too little, education experts told me, to adequately convey the science underpinning the phenomenon. Most of those teachers who did discuss recent climate change in class said they emphasized that global temperatures had risen in the last 150 years. But a third told students that "many scientists believe" the warming is natural, and twice that many encouraged their students to debate between a human or natural cause.

Innocuous as it may seem, teaching climate change as a debate can be as damaging to student learning as outright denial, education experts told me. In general, debate in class is good pedagogy; it helps students pay attention, think critically, and care. It can also convey to students the tentative nature of science—that findings can be overturned by future study. But teaching a well-understood phenomenon as an open empirical question ignores another essential characteristic of science: that it is cumulative. "When a finding is replicated scores of times, hundreds of times, thousands of times, it assumes the status of a fact," Plutzer said. No teacher would encourage a class to debate cell theory, when there is no evidence for a competing theory, and neither should students be asked to debate whether significantly raising the amount of carbon dioxide in the atmosphere does or does not heat the planet. Instead, the National Center for Science Education suggests having students wrangle over questions that real scientists do: competing theories of how much sea levels may rise, say, or how bird migration patterns will change. And certainly, what we should do about global warming—if anything at all—is an open question.

There are a few theories about why educators aren't better at teaching climate science. One is that teachers themselves lack information to teach the subject well. Plutzer's and other surveys have found that a majority of teachers falsely believe scientists still disagree about the cause, which would explain why they teach that. Depending on when and where they went to school, they may not have been taught the subject at all. Climate science fits most squarely under the umbrella of Earth and environmental sciences, neglected branches of school sciences. A federally funded survey of 3,500 science and math teachers found that a third of high schools don't offer those classes even as electives. They aren't very popular classes in those schools that do: While 34 percent of high schoolers take biology each year, just 5 percent take Earth science, and another 5 percent environmental science. Even Earth science teachers were more likely to have taken biology than Earth science in college. This means many science teachers whose job it is to teach climate change never learned about it themselves. Without a solid foundation in the subject, even those with the best intentions can go awry. I once asked an Idahoan teacher whether she taught climate change. "Oh yes! I have a whole lesson on the hole in the ozone layer!" she said, conflating two very different environmental issues.

A second theory is that teachers may feel overt or subtle pressure from parents or colleagues not to teach the subject. "If a teacher thinks climate change is going to be sensitive in their community, chances are they can ditch it," said Plutzer. "They're not expected to cover it at great length. Their students aren't tested on it in a high-stakes test. There's no blowback." There is some support for this theory. The survey of 3,500 teachers asked to what extent "community resistance to the teaching of 'controversial' issues in science (e.g., evolution, climate change)" was a problem in their school. One in five described it as either "somewhat of a problem" or a "serious problem." But that figure isn't high enough to explain why so many teachers miseducate their students about the cause of the climate crisis.

The theory with the strongest support is the simplest. America's teaching force, like its population, spans the political spectrum, and in some cases—take Ms. Sampley's predecessor, for example—a teacher's pedagogy is rooted in a staunchly conservative ideology. An analysis by Plutzer's team found the biggest predictor of how a teacher would approach climate change was their political orientation. "Right-leaning teachers devote somewhat less time to global warming and are much more likely to encourage student debate on the causes," they wrote in 2018. Relevant is that nine out of ten science teachers are white, and over half of science teachers are over forty. (Across all grades and subjects, eight out of ten teachers self-identify as white, while just half of public school students do.) Older white people are more likely to deny the human causes of climate change than young people or people of color.

More education will rarely prevail on a teacher entrenched in their beliefs, thanks to a human phenomenon dubbed "motivated reasoning"—the tendency to seek out information that supports the views we already hold, and ignore information that challenges those views. Children, fortunately, are less susceptible to this phenomenon than adults, and in fact, their views on climate change respond well to a robust science education even in conservative communities. Researchers followed 369 middle-school students in North Carolina as they took climate change units, and found that the students' beliefs around climate change before the class did not predict what they would think afterward. Far more influential were the beliefs of their teachers. This suggests that as long as they get accurate information, some children are willing to learn it, regardless of what their family or community thinks. It also suggests that teachers hold real sway over their students' views of the issue.

But for a child to receive a robust education about recent climate change, she must have a teacher committed to the issue, particularly if she lives somewhere like Oklahoma. In 2015, researchers Nicole Colston and Toni Ann Ivey of Oklahoma State University at Stillwater asked 115 Oklahoma science teachers about their approach to climate change. More than half did not know that scientists are in consensus on the issue. A third believed that recent climate change was mostly natural. And about a quarter said "concern over classroom pushback" kept them from teaching the subject.

Some lucky students, however, get the likes of Melissa Lau, a sixth-grade science teacher in the Oklahoma City suburb of Piedmont, my third school visit in the state. A few years ago,

Ms. Lau participated in a program that matches teachers with field scientists. She traveled to Alaska with a team researching the warming Arctic, where she helped measure the thaw of the tundra and the shift in plant life cycles. She stayed in a village whose coastal roads have already eroded into the oceans. Ever since, she has made a point to teach her students about global warming, regardless of state mandate. I arrived at the end of the school day, just as the class was filing out and Ms. Lau was racing off to bus duty. Three girls stayed behind to tidy up, stacking chairs and picking up colored pencils from the floor. As the three girls color-coordinated the wayward pencils into bins, I asked them if they had heard of global warming before Ms. Lau's class.

"No," said Cadence, a tall girl wearing a purple owl shirt.

"I only heard the name," said Carter, a compact, matter-of-fact kid.

"My grandpa said it was fake," said Claire, whose blond hair was pulled into a ponytail. "I told him we were going to learn about climate change and global warming, and he said, 'Well!'"—here she made her voice sound like an authoritative old man—"'You tell your teacher that your grandpa thinks it's a hoax!'"

"My parents don't believe in it either," Carter volunteered.

"Do you?" I asked them.

"I'm not sure," Claire said, hesitating. "I was like, how could it be fake or real? And then we learned about it and I was like, well, there's a bunch of data pointing to it, so . . ."

I asked them if they thought global warming might affect their lives at some point.

"No," said Carter.

"I don't know how it would," said Claire.

"It could possibly affect us, but part of me thinks it won't," said Cadence.

The three girls left for their after-school business club, where they would be preparing for a field trip to a mall in Tulsa. When Ms. Lau came back from bus duty, I told her what the girls had said.

"It's really difficult for kids like Claire." She sighed. "They go home and the most significant people in their lives—their parents, their grandparents—are saying it's not real. As an educator, how do I say, *well, your parents are wrong*? I can't say that."

Izerman's parents had a long list of reasons to consider moving to Oklahoma. Climate change was on it, as were health care and more job opportunities. Probably the biggest, though, was chasing a better education for their children. Izerman's father explained that he meant no disrespect toward the island's teachers; he had been educated by them himself, and saw the pains they took to do a good job. But he wondered if US schools would have more resources to offer Izerman and his siblings. The move wouldn't need to be permanent for that. Just long enough, he said, "for our kids to get a really good schooling."

The Evolution

It's just not in Barbara Forrest's DNA (as she is wont to say) to sit back and let clowns come in and screw up the public school system. So, in 1994, when three men from the big city came to her quiet Louisiana parish and convinced the local Christian Coalition to push a hogwash curriculum, she refused.

For more than fifty years, Forrest has lived on a farm in Livingston Parish, the pine- and cypress-forested terrain between Baton Rouge and Lake Pontchartrain. Her husband was born and raised in the parish, and she moved there when she was sixteen. The couple ran a poultry farm on his family's land to pay their way through college and then graduate school, and were glad to raise their children there as well. "We're a rural area, we're not a rich parish, but we have a very good public school system. A no-nonsense school system where you send your kids and they're gonna learn stuff."

One day in May 1994, three members of a New Orleans group called the Origins Resource Association visited Livingston Parish to attend a meeting of the local Christian Coalition;

they were there to discuss "policy on the teaching of evolution," according to a meeting agenda Forrest later obtained. The discourse was apparently productive, because soon, the Livingston Parish school board received a petition from the Christian Coalition signed by 1,500 residents asking the board to put "intelligent design" into science classes as an alternative to evolution. They also received a proposed new curriculum guide written by the Origins Resource Association. The curriculum would, among other things, ask students to "list evidence for an old Earth and a young Earth," and explain how an esoteric paper by a Nobel Prize laureate supported the idea that the second law of thermodynamics could temporarily be overridden. The district handed the question over to a twenty-five-person science curriculum committee made up of local teachers. That they were even considering it troubled Forrest, whose son was a high-school freshman. "I wasn't about to sit back and let these guys come in from outside to do what they want," she said. "It's just not in my DNA."

So Forrest spoke out against the curriculum in meeting after meeting. She wrote to dozens of scientists, many of whom responded with letters to the school board decrying the curriculum guide. The Nobel Prize winner himself wrote back, asserting that his theories "do not in any way find an exception to the second law of thermodynamics." Forrest, a philosophy professor at Southeastern Louisiana University, enlisted a colleague in the biology department to help dissect the curriculum in search of weaknesses, and sent the resulting analysis to every member of the committee and school board. Some were more receptive than others. When she made an appointment to visit her local school board representative to discuss the issue, he drove his tractor out to the field before she arrived, standing her up.

The curriculum committee voted 23 to 2 to reject the proposal. But rather than abide by the committee's recommendation, the school board continued to consider adopting the curriculum. Forrest alerted the American Civil Liberties Union, who wrote the board promising a lawsuit if it approved the materials. So a board member sympathetic to the Christian Coalition's cause introduced a substitute proposal: If a student brought up a non-evolution theory, it could be incorporated into "that day's teaching plan." That proposal passed 5 to 4. Nonetheless, Forrest felt a thrill of success at defeating the curriculum. The three "clowns" from the New Orleans organization "packed up their bags and left town," said Forrest. She had little way to know that this first bout with anti-evolutionists would not be her last.

Long before climate change was a household word, educators were fighting off attacks on another topic: evolution. In 1859, when Charles Darwin published *On the Origin of Species* introducing evolution by natural selection, few Americans paid attention. But as the theory gained evidence, it began producing answers to certain questions that had previously belonged in religion's domain. Church leaders with a strict interpretation of the Bible spoke out against the threat they saw in Darwinism. People like three-time presidential candidate William Jennings Bryan believed that teaching children about evolution would threaten the nation's moral compass. "If one actually thinks that man dies as the brute dies, he will yield more easily to the temptation to do injustice to his neighbor," he said in a 1904 lecture.

In 1923, Oklahoma adopted a bill prohibiting the purchase of textbooks that included the Darwinian theory, and Bryan persuaded the Florida legislature to pass a nonbinding resolution advising educators to avoid teaching "any theory that relates man in blood relationship with any lower animal." Finally, in March 1925, a state adopted a complete ban on teaching human evolution in school: Tennessee's Butler Act. Immediately, the American Civil Liberties Union announced a search for a Tennessean teacher willing to test the law in court. Some folks hanging out at a drugstore in the Appalachian town of Dayton, population 1,800, hypothesized that having the trial in their town might bring in tourists. They sent a kid to fetch a science teacher named John Scopes from a nearby tennis court. With some persuasion, he agreed to volunteer for the case. Within days he was formally arrested and accused of teaching evolution. Bryan offered to help the prosecution, and Clarence Darrow, then America's most famous defense attorney, assisted the defense. At the end of the so-called "Monkey Trial," which captured headlines worldwide, the jury took nine minutes to convict Scopes. The judge fined him $100. The Tennessee Supreme Court denied an appeal, but also overturned the conviction on a technicality, quashing the case.

The Monkey Trial reinforced the burgeoning belief among Americans that science and religion were in opposition—a notion that has threaded through American culture and politics ever since. It also marked the first major skirmish in a century-long war over what children should be taught about evolution. As Christian fundamentalists grew in numbers and strength, these skirmishes would play out hundreds of times, statehouse by statehouse, district by district. The question finally found itself back in court forty years after the Scopes Trial. Little Rock biology teacher Susan Epperson had sued Arkansas, arguing that the state's ban on teaching evolution kept her from fulfilling her duties as a "responsible teacher of biology." In November 1968, the US Supreme Court ruled that evolution bans violated the First Amendment's Establishment Clause, which prohibits the government from adopting or favoring a religion.

The anti-evolution movement digested the loss, but did not retreat in defeat. Instead, for the first of at least three times, the campaign reinvented itself in hopes of achieving its goals within the confines of the Establishment Clause. The movement's first makeover replaced creationism with the concept of "creation science," which framed the divine creation narrative in scientific terms—though, when pressed, its proponents admitted there was no way to test the so-called science, since it depended on God performing miracles. The school districts of Dallas, Chicago, and Atlanta mandated that their science teachers present creation science alongside evolution, and lawmakers in dozens of states introduced "balanced treatment" bills. As they did, they met opposition from a growing pro-science grassroots network. In 1981, an organization called the National Center for Science Education was born with the purpose of tracking anti-evolution efforts and coordinating resistance to them. By then, the ranks of anti-evolutionists had grown as well, as conservative Christian leaders threw in with the Republican Party and shaped their flock into a political force. It was obvious the courts would need to weigh in once again. When a "balanced treatment" bill passed in Louisiana in 1981, a high-school teacher at a district ninety minutes west of Barbara Forrest's parish led a suit against the state. When the case reached the Supreme Court in 1987, the court ruled that creation science was just religion in a lab coat, and that teaching it in school was unconstitutional.

A new idea emerged: intelligent design. The concept was popularized by the 1989 book *Of Pandas and People: The Central Question of Biological Origins*, later promoted by a Seattle think tank called the Discovery Institute. The book avoided explicit reference to the biblical origin story, and conceded that *microevolution* occurs—one might breed a sweeter variety of corn, for instance. However, it argued that *macroevolution*, in which one species evolves into another, is a mere hypothesis. Instead, it proposed, an unnamed "intelligent agency" caused forms of life to appear abruptly and intact. Groups like the Origins Resource Association showed up in places like Livingston Parish to push alternative curricula based on the idea. The extent of the movement's ambition was revealed in a Discovery Institute fundraising document, which framed the issue in moral terms: The proposition "that human beings were created in the image of God" had suffered "wholesale attack by intellectuals" who portrayed humans "not as moral and spiritual beings, but as animals or machines." The document laid out a strategy it said could make intelligent design the dominant perspective in society within twenty years. Barbara Forrest, fresh off her battle in Livingston, delved into an analysis of the document. In 2004, she and scientist Paul R. Gross published a book connecting the dots between the ostensibly nonreligious idea of intelligent design and the religious lineage of creationism.

The concept was ripe for another test in court, and it found that test in the Conewago hills of central Pennsylvania. The board of the Dover Area School District, which served a community of about 20,000, had ordered that their students be introduced to intelligent design. Eleven parents sued the district. During discovery, the legal team subpoenaed the creators of the canonical *Of Pandas and People* and received five early drafts of the book—four written before the Supreme Court banned creation science in 1987, and one after. The "before" versions included the word *creationists* where the "after" version said *design proponents*. Forrest, who helped with the team's analysis, found one case where someone had accidentally only half-erased *creationists*, for a result of "*cdesign proponentsists.*" There could be no clearer evidence that intelligent design merely gave a new name to an outlawed practice. The judge ruled that teaching intelligent design in public school, like its predecessors, violated the Establishment Clause. The decision was met with so much fury from anti-evolutionists that the judge's home had to be guarded by the US Marshals, and televangelist Pat Robertson said that if a disaster were to hit Dover, its citizens should not look to God for help.

The anti-evolutionists had evolved twice—first from creationism to creation science, then to intelligent design. Now another adaptation was necessary. "We have entered a new front in the debate over intelligent design—the need to protect academic freedom," announced a Discovery Institute newsletter published just weeks after the Dover defeat. That term—*academic freedom*—harked back to a bill called the Academic Freedom Act proposed in Alabama two years earlier, which would protect teachers from disciplinary actions if they presented "information pertaining to the full range of scientific views concerning biological or physical origins" in class. "Nothing in this act shall be construed as promoting any religious doctrine," the bill stated. It didn't get the votes to pass Alabama's legislature in 2004, but it lit a spark. The Discovery Institute created a model academic freedom bill and made it available to a network of sympathetic lawmakers around the country. Proposals were put forth in Oklahoma, Mississippi, Maryland, New Mexico, and Florida. Those failed, but in Mississippi in 2008, similar language was tacked onto an unrelated bill and passed, allowing teachers there to discuss the "full range of scientific views" about the origin of life without fear of repercussions.

I interviewed Discovery Institute's vice president and senior fellow John West about the group's model legislation in 2017. "This isn't about the Bible or biblical literalism," he said. "It doesn't protect teaching about intelligent design, let alone about creationism." Rather, the bill is meant to protect against a "dogmatic and narrow" teaching of evolution, and provide a "safe haven" for teachers who present evidence that "raise some questions about what natural selection is good at doing and what it's not good at doing." Scientific consensus "should be taught first and foremost," he said, but one must always be evaluating and questioning the evidence. The scientific community, however, saw the model bill as yet another attempt to smuggle creationism into the classroom. "Academic freedom is a euphemism," Forrest told me. "Laws about science education don't need religion disclaimers."

Since then, one more evolution has occurred: the cross-pollination of "academic freedom" and climate change. In 2006, teachers in Louisiana's Ouachita Parish—population 150,000, about four hours northwest of Forrest's home—asked for clarification of their rights to teach non-Darwinian theories. In response, their school board adopted an academic freedom policy, with one crucial amendment. In addition to allowing teachers to present the "scientific strengths and weaknesses" of issues important to the Christian Right—evolution, the origins of life, and human cloning—it also specifically protected questioning of global warming. The next year, Ouachita School District's assistant superintendent ran for the Louisiana state legislature and won. After he took office, he sponsored a bill that would expand the Ouachita policy statewide—including the explicit global warming reference.

Forrest once again found herself campaigning to protect science education. But this one she lost. Louisiana's House passed the bill 94 to 3, the Senate did so unanimously, and Governor Bobby Jindal signed it into law. Since then, legislative efforts targeting both evolution and climate change in science class have been proposed in Arizona, Colorado, Florida, Indiana, Iowa, Kentucky, Michigan, New Mexico, Oklahoma, South Carolina, South Dakota, Texas, and Virginia. Those efforts have failed to garner enough votes to pass. But in Alabama and Indiana, lawmakers adopted nonbinding resolutions "urging" support in schools for diverse views on the subjects. And in 2012, nearly nine decades after Scopes's prosecution, an academic freedom act received enthusiastic support from the Tennessee legislature and became law. Nicholas Matzke, a former NCSE staffer who worked on the Dover case, has traced the evolution of the language used in academic freedom bills like Tennessee's. He found Ouachita's innovative inclusion of global warming has become law governing more than 11 million people.

In effect, the Ouachita policy formally merged the interests of anti-evolutionists with those of climate denialists. Though the two groups are motivated by strikingly different things—one by libertarian economics and the other by interpretation of scripture—they are now married in the conservative mind. These days, the two show up as a package deal in almost every fight over science education, whether that fight is over textbooks, academic standards, or legislation.

They were never inevitable allies. Libertarians tend to be the least religious contingent of the Republican coalition, and many Christian denominations have been tugged by a "creation care" movement toward the belief that people must be good stewards of the earth. In most cases a person's religious beliefs do not interact much with their thoughts on global warming. A 2016 study found that, as a rule, someone's political views matter far more than their religious views in predicting their opinion on the climate crisis. For instance, if you compare two equally conservative individuals, one religious and one not, they are equally likely to be a climate doubter. There's a major exception to this rule, however: Evangelicals are considerably more climate skeptical than their politics would predict.*

For a while, evangelicals seemed to drift away from climate denial instead of toward it. Their highest political priorities have long been "below the belt" issues like abortion rights and

* Categorizing the religious diversity of Protestants is a complicated affair, but most researchers divide them into three major groups: *mainline Protestants* from historically white denominations, *evangelical Protestants* from historically white denominations, and *Black Protestants* from historically Black denominations (which, whether mainline or evangelical, have markedly different politics from their historically white counterparts). For the purposes of this chapter, "evangelicals" refers to members of historically white denominations that trace their roots to the fundamentalist movement, adhering to what they see as the "fundamentals" of the faith (the authority of Scripture, the veracity of the miracles described in the Bible, etc.).

same-sex marriage, but creation care proponents have sometimes rallied their congregations' support behind environmental issues like endangered species and water pollution. In the early 2000s, several environmentalist groups and funders reached out to evangelical faith groups and offered to bankroll an effort to grow support for climate action. Several major grants went to an organization called the Evangelical Environmental Network, which started building a "bench" of church leaders willing to present climate action as a moral imperative. After several years of mustering support, the group launched the Evangelical Climate Initiative in February 2006. The initiative's founding document was a call to action signed by eighty-six church leaders. It made such a splash that, within months, its leader, Richard Cizik, was photographed walking on water for the cover story of *Vanity Fair*'s "Green Issue." In the story, Cizik quoted chapter 11, verse 18 of Revelation, which warns that God will "destroy those who destroy the Earth." Hope soared among environmentalists that the evangelical community might embrace the reality of the climate crisis.

Then came the backlash. Since the 1970s, the politicized arm of the evangelical tradition, often referred to as the Christian Right, has worked to unify evangelicals around a conservative political agenda. By President Ronald Reagan's first election in 1980, they had become an "anchor group" of the Republican Party, a role that grants them both power and responsibility, as detailed by scholars Lydia Bean and Steve Teles in their 2015 report, *Spreading the Gospel of Climate Change*. In exchange for broad Republican support for issues important to their churches, evangelical leaders have spent decades convincing their flock that the Republican ideal of small government is the Bible's take on modern politics—effectively working libertarian ideals into the psyche of their congregants. These church leaders are also expected to "police unorthodoxy within their own ranks" when it threatens another anchor group in the party. One of those other anchor groups is the fossil fuel industry, on whom measures like carbon dioxide regulation would impose a large cost.

So the policing began, Bean and Teles write. As soon as the Evangelical Climate Initiative launched, Christian Right leaders began pressuring signatories to remove their support. Many did. When several Southern Baptist leaders signed a separate letter voicing concern about the climate crisis, they received a call from an evangelical organization in Washington, DC, threatening to send an email "to every Southern Baptist" directly challenging their authority to speak for the church. The conservative evangelical organization now known as the Cornwall Alliance roared to life, calling on the National Association of Evangelicals to refrain from taking a position on climate. Several well-known leaders of the Christian Right wrote another letter asking for the resignation of Richard Cizik, whom they accused of "using the global warming controversy to shift the emphasis away from the great moral issues of our time," such as sexual abstinence and same-sex marriage.

Suddenly, global warming became a subject of great interest in the vast and powerful evangelical media world, which, while invisible to many secular people, reaches 90 percent of evangelicals and about one in five Americans each month. It did not, however, become a subject of great debate. Scholar Robin

Globus Veldman, author of the book *The Gospel of Climate Skepticism*, reviewed climate coverage on evangelical radio, television, and online outlets between 2006 and 2015 and found it "heavily tilted toward skepticism." Evangelical radio pundits Chuck Colson, Ken Ham, Janet Parshall, and Tony Perkins, among others, collectively talked about global warming scores of times on shows syndicated to as many as 1,200 stations, and Veldman did not find a single instance where they acknowledged the reality of the climate crisis. Televangelist Pat Robertson flip-flopped on the subject, but the matter was presented skeptically forty-six of the fifty-eight times it came up on his TV shows, which had an audience of 1 million daily. At least three times before he died in 2007, Rev. Jerry Falwell spoke skeptically of the climate crisis on his television show, *The Old-Time Gospel Hour*, which reached as many as 12 million people.

These media ministries presented global warming alongside issues like abortion, prayer in schools, and creationism, transforming it into a religious issue and conveying the sense that the biblical view on the climate crisis is that it's not happening, Veldman found. Progressive evangelicals lacked their own media network to get their messages out, so denialist messages overwhelmed all others, creating a bulwark against any other takes on the issue. And because of evangelicals' long history of standing with the Republican coalition, the libertarian messaging wasn't a hard sell.

It's unclear exactly how coordinated this campaign was, Veldman says—whether it entailed evangelical leaders sitting around a table and agreeing to spread climate denialism, or something looser. "All I can say is that a bunch of different individuals who were all broadly part of the Christian Right began to talk about climate change at about the same time in about the same way," she said. Their efforts worked. In 2006, only 25 percent of white evangelicals disbelieved that climate change was occurring. By 2014, that figure was 39 percent. Those who conceded the climate was changing but didn't believe humans had caused that change grew even more precipitously. Meanwhile, the rest of the public moved in the opposite direction.

If a conversation around a table did happen, it was likely attended by the Cornwall Alliance, with which most of the denialist evangelical pundits have been affiliated. The group specializes in adapting secular climate denial talking points for an evangelical audience. Its founder, E. Calvin Beisner, has testified before both houses of the US Congress, traveled the Christian college speaking circuit, and frequented the evangelical media scene. In 2010, Beisner appeared on Glenn Beck's show on Fox News to promote a package of congregational resources called "Resisting the Green Dragon" that presents the environmental movement as "one of the greatest deceptions of our day." Imagine an architect who designed buildings that collapsed if you leaned on the wall, he told Beck. "We'd all say, that was not a wise architect, right? Why are we saying that the God who designed this creation made the climate system so that a tiny change in atmospheric chemistry could send all—all of it to catastrophe?"

Beisner's path has also crossed that of the Discovery Institute, namely in 2015 when he delivered a "lunch talk" there titled "Climate Change and the Poor." The institute's John West told me that while he's "vaguely aware of some of the groups" that work to spread climate denial, his organization has not coordinated with them nor received funding from the fossil fuel industry. Be that as it may, the website *DeSmog* has uncovered several forays the

Discovery Institute has made into the climate denial ecosystem. In 2006, Discovery hosted a talk by a coauthor of the denialist classic "Unstoppable Global Warming: Every 1,500 Years." The next year, Discovery's cofounder introduced a conference panel titled "The Global Warming Myth." More recently, Discovery senior fellow Jay Richards appeared on denial think tank Heartland Institute's podcast to discuss how climate action is "unnecessary." And in October 2020, Richards and another Discovery fellow paired up with a Heartland policy advisor to write a book questioning a different scientific subject: *The Price of Panic: How the Tyranny of Experts Turned a Pandemic into a Catastrophe*.

Ouachita Parish School Board's matchmaking between the evolution and climate change "controversies" occurred at the start of the evangelical campaign against climate action. But there may have been another reason to bring climate denial into the creationism fold. Academic freedom policies have never been challenged in court, their vague language offering enough plausible deniability that such a challenge could be risky. By merging the forces of religious fundamentalism and free-market fundamentalism, the Ouachita innovation offers another layer of defense were a legal challenge to arise, because while separation of church and state is enshrined in the Bill of Rights, no such clause separates business interests and state. To Barbara Forrest, it's obvious that the two strands of activism came together through mutual interest. "The Religious Right realized they could hitch their wagon to a powerful new current—climate denial—which has amplified their influence. And the anti-climate-science people got a bunch of foot soldiers they didn't have to pay for."

Every year now, lawmakers in states across the country propose bills that aim to influence how evolution and climate change are taught in science classes. Most are academic freedom acts, which have been introduced at least seventy-five times in twenty states since the Discovery Institute announced academic freedom as the "new front in the debate over intelligent design." That they have not found more success—only Mississippi, Tennessee, and Louisiana have adopted them—is largely thanks to the work of the NCSE, which tracks the bills and galvanizes local opposition to them as they arise. Though the NCSE was established in 1981 to defend the teaching of evolution, in 2012, pushback on climate education had become so common that it expanded its mission to defend that as well.

Barbara Forrest feels extraordinarily proud of her role in championing science education over the years. Just as anti-evolutionists like William Jennings Bryan and the Discovery Institute have presented themselves as crusaders for American morality, so does Forrest. To her, allowing creationism into science classrooms would foster a world where evidence and fact have no real weight. A survey of 1,427 teachers by Plutzer and the NCSE in 2019 found educators more willing to present evolution in class than there used to be, but nonetheless, 6 percent of biology teachers said they primarily emphasized creationism, 12 percent said they presented creationism alongside evolution, and 15 percent avoided endorsing either evolution or creationism as scientifically credible. Forrest can expect more work to uphold her version of American morality ahead.

The Standards

Had Izerman moved from the Marshall Islands to Oklahoma, he might have learned something in high school about the phenomenon overwhelming his homeland—*if* the school offered elective Earth or environmental science classes, and *if* he chose to take them. But were his family to move to Hawaii instead, he could have learned about the climate crisis in third-grade social studies, middle-school science, and high-school biology, US history and government, world history and culture, Pacific Island studies, Earth science, environmental science, and at least one math class.

The difference amounts to academic standards, a state's expectation of the knowledge and skills learners will master in each grade and subject. In contrast to a curriculum—the lesson plans, lab kits, and other resources that shape *how* a subject is taught—standards shape *what* is taught. A public school's curriculum, along with its textbooks and tests, are heavily informed by the standards its state has adopted. Mundane as they are, standards represent a state's greatest level of control over what children learn. That Oklahoma's made no mention of recent climate change was not accidental. To the contrary, the absence had been engineered.

Once upon a time, most states left it up to schools to decide what to teach kids about science. Some states eventually began providing recommendations for instruction in each grade, as Oklahoma did as early as 1960. But mandatory statewide standards as we know them today weren't born until the 1980s, when the Reagan administration, citing "the widespread public perception that something is seriously remiss in our educational system," created a national commission to study the problem and dispense advice. The commission recommended "rigorous and measurable standards." States like Oklahoma put together panels of educators to consider what those standards should include. Once written, the documents went to state legislatures for a thumbs-up, and then became law. Oklahoma put its first science standards into place in 1993 and has revised them about every six years since.

America loves to fight over what happens in classrooms, so the seemingly bureaucratic exercise of updating academic standards has often taken on a caustic political life. This was true even before the Bush administration's No Child Left Behind Act tied federal funding for schools to high-stakes tests in reading and math, and tied those tests to state standards. After that, they were on the radar of every education advocate. In the sciences, naturally, standards' treatment of evolution became a keenly negotiated question. When Ohio adopted new science standards in 2002, they required students to "describe how scientists continue to investigate and critically analyze aspects of evolutionary theory." The policy, praised by the Discovery Institute and pilloried by science educators, found a home in the standards of seven other states within a few years.

NGSS

The typical state science standards at the time were basically lists of information and skills children should know by the end of a given year, but pedagogical researchers had amassed evidence that simply asking children to memorize Newton's second law of motion, say, would not train them to be physicists. Instead, the experts said, students benefited from *thinking* like scientists: gathering evidence, spotting patterns, inventing contraptions, reasoning through cause and effect, devising hypotheses, and experimenting on the world around them.

In 2012, the National Research Council released the "Framework for K–12 Science Education," which laid out this more holistic vision of what should happen in a science class. A team of educators then used it as the basis for the Next Generation Science Standards, a set of model standards for states looking for a paradigm shift. Under the NGSS, teachers who once had their students commit Newton's second law of motion to memory might instead hand them a skateboard, a timer, and a measuring tape so they could investigate for themselves the relationship between force, mass, and acceleration.

The model standards took an unapologetic approach to evolution, and they also embraced modern climate science. Some states had already begun to tiptoe toward covering global warming. By 2008, thirty states' science standards mentioned some aspect of the phenomenon, though almost always in an elective class and rarely with detail or clarity. In contrast, the NGSS put it right in middle-school science, which every student had to take, and laid it out plain:

> Human activities, such as the release of greenhouse gases from burning fossil fuels, are major factors in the current rise in Earth's mean surface temperature (global warming). Reducing the level of climate change and reducing human vulnerability to whatever climate changes do occur depend on the understanding of climate science, engineering capabilities, and other kinds of knowledge, such as understanding of human behavior and on applying that knowledge wisely in decisions and activities.

As soon as the model standards were published in April 2013, the educators who had helped write them went to their state capitols to advocate for them. The very next month, Rhode Island adopted NGSS. Then came swift adoptions in Kansas, Maryland, and Vermont. That fall, the NGSS ran into trouble in Kentucky, where a conservative lawmaker wrote an op-ed decrying their treatment of climate change and evolution and warning of "political correctness" becoming "the arbiter of learning." A legislative committee voted 5–1 to kill the standards. However, the governor managed to make a procedural end-run around the legislature and push the standards into schools without its approval.

The next year, Oklahoma's science standards came up for revision, so the state's education department appointed a committee of fifty-nine educators to consider updates. Adopting anything like NGSS would be tremendous progress for Oklahoman science education, long guided by standards so flawed that they had received an "F" grade in an assesment by the Fordham Institute. The committee recommended that the state adopt a close replica of the NGSS—with some "tweaks." Among other changes, they replaced every instance of the word "evolution" with "change over time" and other euphemisms that had always stood in Oklahoma standards. Committee member Julie Angle, an educator who spent twenty-five years teaching high school and a decade training pre-service science teachers at Oklahoma State University, told me she opposed the "tweaks" but understood why she got nowhere. The committee feared the legislature would scrap the entire proposal, so they preemptively forfeited the controversial bits. "I don't think Oklahoma was ready for the 'e-word,'" Angle said. "We have legislators who believe that evolution has not, is not, and will never occur in the great state of Oklahoma." The committee also axed all mention of human-caused climate change from required courses. Even in elective Earth and environmental science classes, discussion of the phenomenon was pruned down.

The NGSS-lite standards had two stops to make. First stop, the Oklahoma Board of Education, where they met with approval. Second stop, the state legislature, where they met no such luck.

On a blustery Monday in May 2014, the Oklahoma House Administrative Rules committee flared over the proposal. Some lawmakers despised the idea of taking cues from a national model—another example of federal meddling in local education, they said. They also fretted that even stripped of climate change, the standards could somehow be used by teachers who low-key really wanted to teach it. In early grades, children would be learning about weather and seasons, and how different parts of the world experience different climates. Could those lessons, Republican state Rep. Mark McCullough asked, "potentially be utilized to inculcate into some pretty young, impressionable minds . . . a fairly one-sided view as to that controversial subject, a subject that's very much in dispute among even the academics?" McCullough's committee recommended 10–1 that the legislature reject the proposed standards. Oklahoma's House of Representatives took that recommendation, voting 55–31 on a resolution to reject. The proposal's woes continued in the state Senate. Saying that "global warming is the main concern," state senator Anthony Sykes tacked their rejection onto another bill, which passed.

Fortunately, under the rules governing Oklahoma's legislature, the standards could only be killed if the House and Senate voted to reject the *same resolution*. The two bodies had used two different pieces of legislation to do so, and their session ended before they could fix that procedural error. Oklahoma's watered-down version of NGSS passed into law and was soon guiding lessons in science classrooms across the state—much improved over the last version, but censored nonetheless.

As more states bought into the Next Generation standards, more states battled them. Wyoming's legislature added a footnote to their state budget precluding the use of state funds for "any review or adoption of the NGSS"; educators successfully advocated for its repeal the next year and pushed a set of NGSS-style standards through. When the standards survived an initial dispute in Iowa, conservative lawmakers proposed legislation to repeal them, which failed, and then to block their implementation, which also failed. Louisiana approved standards based on the same framework as NGSS but in an accompanying bulletin reminded educators about the Louisiana Science Education Act—the law allowing science teachers to raise alternative views about evolution, origins of life, human cloning, and climate change without fear of discipline. In West Virginia, the state board of education tampered with their committee's recommended standards, also based on the NGSS framework, to weaken their treatment of climate science; sustained public outcry led to a compromise. In New Mexico, the education department removed material about evolution and climate change, presumably to avoid criticism, but instead faced so much disapproval that they changed it back.

The most dogged fight took place in Idaho. In 2016, the state legislature sent a set of NGSS-style standards back to the committee of educators who had recommended them. In 2017, the committee came back with a modified version, but stood their ground on the inclusion of climate change. This time, the legislature blocked *just the five standards* that referred to climate change. In 2018, the educators remained stalwart that Idaho children should learn climate science, and brought back the climate standards a third time. Idaho's House again voted to reject, but by then the issue had garnered national attention, and the state Senate caved to public pressure and allowed the climate standards to stand. The issue finally seemed to be at rest, but in 2019, Idaho's House pulled an unprecedented political maneuver that brought the standards into question again. The state was forced to hold public hearings on the subject a fourth year in a row. The standards held.

Despite all the wrangling, the nation's antiquated science standards have largely yielded to an embrace of the NGSS's paradigm-shifting approach. As of 2021, twenty states and the District of Columbia had adopted the NGSS whole cloth. Twenty-four other states had adopted a modified version, though many included "tweaks" like Oklahoma's. The remaining six states stuck to their own standards. Because those six include populous states like Texas, Ohio, and Florida, they represent 29 percent of the nation's student body.

We know standards are influential: In a 2018 survey, 84 percent of middle- and high-school science teachers agreed with the statement that "most teachers" in their school "teach to the state standards." We also know that what standards say about climate change makes a difference: Teachers in states whose standards include *anything* about climate change spend significantly more time on the topic than those whose state standards avoid the subject. Lastly, we know that as lawmakers in some red states have worked to shrink what children learn about the climate crisis, lawmakers in some blue states have worked to expand it.

In 1987, Frank Niepold was meandering through a bookstore near the art school he attended in Halifax, Nova Scotia, when a book about the earth caught his eye. The planet had been making regular appearances in the art he made, so he pulled it out and leafed through until he came to a colorful, double-page spread about the greenhouse effect and how humans had intensified it, to potentially calamitous effect. "I stared at it for a really long time," he told me. "It got in my brain that there's something going on with the earth that is important and requires my focus." From then on, his art focused exclusively on the earth and its impending crisis. After graduating, he felt limited by his medium, so he became a human ecologist, and then an Earth science teacher. Eventually he landed a job at NASA working as an education specialist as part of the agency's "Mission: Planet Earth." That was satisfying for a while, but then, he said, "NASA got distracted with all their moon and Mars nonsense. I kept saying to them, 'That's just a distraction. We need to focus on the home planet.' And they were like, 'Moon! Mars!' And I was like, 'Home planet!' And they said, 'Moon! Mars!' So I said, 'I'm going to lose this argument, so I need to find a new agency.'"

He asked himself what agency would *never* lose focus on the home planet, and the answer to him was obvious: The National Oceanic and Atmospheric Administration, a scientific agency under the US Department of Commerce. "At NOAA, climate is our North Star. Climate is our bones. Climate is our mitochondria. It's everywhere," he said. Even when the Trump administration forced the Environmental Protection Agency to remove climate education materials from its website, NOAA's stayed put. As NOAA's climate education czar, Niepold oversaw the development of educational materials, traveled to hundreds of education conferences promoting them, participated in youth delegations to UN climate conferences, and partnered with any visionary climate education project that would have him. Niepold is known as such a zealous promoter of climate education that several weeks into the Biden administration, Special Envoy for Climate John Kerry nominated him to help lead America's international collaborations associated with Article 12 of the Paris Agreement on Climate Change. "What's Article 12, you might be wondering?" he asked me by Zoom, literally bouncing at his standing desk with excitement. "Article 12 says that signatories are to advance education, training, public awareness, public participation, public access to information, and international collaboration to accelerate the nation's climate action." Frank Niepold has ascended from the nation's most enthusiastic booster of climate education to perhaps its most powerful one.

Since Niepold has now spent decades thinking about what it takes to give American students a useful education about the climate crisis, I asked him what he thought about the Next Generation Science Standards, which he played a role in developing as a community reviewer. "You know . . ." he said, his voice trailing into hesitation. "Look, the NGSS is 70 percent awesome. They're a massive improvement over what we had. They just don't go far enough." Under NGSS, climate change shows up in middle-school science, but it gets scant mention elsewhere, so it's not reinforced. "What's happening is that one teacher might talk about climate change in middle-school science, then two years later it might come up in, like, a humanities class. And the number of students who happen to go through both classes might be, say, 20 percent. So it's like Swiss cheese. The chance of missing things is significant."

A Swiss cheese climate education damages our hope of addressing the crisis, since "there are so many kids out there that just don't have a clue what's happening," he said. "Their brilliance isn't going to be involved in solving this thing." It also does a disservice to students' career prospects, he said. Sectors like agriculture, transportation, and fuel production are undergoing a transformation, and they will need a labor force ready to take on those changes. This will create bountiful opportunities for young people who know enough to capitalize on them. "I constantly hear stories of students who first learn about climate change in their third year in college. They've been thinking the world is operating one way for twenty years—all their internships, majors, college choices are predicated on that idea—and then they're like, 'Wait a minute. You mean to tell me this is real?'" Their education system let them down, Niepold said.

To make a truly climate literate student body, he said, climate should be incorporated in developmentally appropriate ways throughout elementary school. It should be all over biology, chemistry, and physics standards, each of which has clear scientific connections to the phenomenon. And it shouldn't just show up in science. Arts, language, history, civics, and economics teachers all have a role to play in preparing their students for the crisis the world is being transmogrified by, he said. A truly robust set of standards could make this happen

By no means is Niepold alone in this thinking. In 2018, lawmakers in Connecticut floated a bill that would require public schools to teach students about climate change. Democratic lawmakers in other states followed. In 2020, pro-climate-change instruction measures were introduced in twelve states, nine of them blue. In June 2020, New Jersey's board of education inserted mention of climate change throughout science and social studies standards, along with health and physical education, computer science and design thinking, visual and performing arts, world languages, English language arts, and mathematics. This kind of progress fills people like Frank Niepold with hope. But it's doing something darker, too. It's seeding a two-tier system.

Red State, Blue State

In 2020, the National Center for Science Education teamed up with the progressive nonprofit Texas Freedom Network Education Fund to grade each state's science standards based on how they treat climate change. A panel of reviewers assessed how well they convey four essential points of scientific consensus: It's real, it's us, it's bad, and there's hope. The reviewer's responses were assigned numerical scores and collectively converted into a grade. The twenty states and District of Columbia that have adopted NGSS wholesale got a B+, a nod to their room for improvement. Six states did as well or better. Twenty-four states did worse. Mississippi, whose science standards ask middle schoolers to "engage in scientific argument" over "whether climate change happens naturally or is being accelerated through the influence of man" got a C. South Dakota, whose standards acknowledge the "particular sensitivity to two issues: climate change and evolution," got a C−. Six states got Fs.

I enlisted the help of students at Marquette University, where I spent nine months as a public service journalism fellow, to compare state standards to a red-blue-purple map of the states, colored based on whether the state's legislature had been controlled by Republicans, Democrats, or had split control in the years since NGSS was released. It turned out that some of the best grades in the country were red states: Wyoming, North Dakota, and Alaska all got As or A-minuses. But they were not the norm. Of twenty-eight red states, twenty-two had watered-down climate language in their science standards. So did two of the seven purple states. But zero of the fifteen blue states or DC did worse than a B+.

Some states mention recent climate change in civics, history, economics, or geography classes, so my students and I dissected those standards to see what they said. We found the subject completely absent from thirty states' social studies standards. Of those that do mention it, just seven require it, while the rest list global warming as an optional discussion topic. The partisan effect was discernible here, too: 56 percent of blue states' social studies standards included some mention of climate change. For red states, that figure was a more modest 39 percent.

The steady advance of many blue and red states in opposite directions is creating a crude two-tier system, in which children in some places are required by law to learn about the phenomenon that will shape their century, while in others, students may not hear the words "climate change" in class at all. A survey of 832 middle- and high-school science teachers, roughly half from California and half from Texas, found that California teachers had a better understanding of global warming's mechanisms, and Texans more often suggested to students that global warming could be natural. This might just be chalked up to political orientation—more Texan teachers are politically conservative—but even controlling for political affiliation, the Texans behaved differently than their blue-state counterparts. Standards may help explain why one can roughly predict what a child might learn about climate change simply by knowing whether they happen to live in a state where climate change science is broadly accepted, like Hawaii, versus one where it is a source of friction, like Oklahoma.

Six years after Oklahoman educators managed to squeak NGSS-lite through their legislature on the back of a procedural error, their science standards were once again up for revision. This time, the writing committee recommended explicitly using the "e-word"—evolution—rather than the old euphemisms, and reinserting climate change where it had been stripped six years earlier. Their proposal wasn't quite true to the NGSS—it omitted a paragraph describing specific human activities that have altered the atmosphere (namely fossil fuels and agriculture)—but the fundamentals were there.

Like its predecessor, this proposal had two stops to make. In February 2020, I traveled to Oklahoma City to watch its first stop at the state's education board. The meeting was held in a conference room on the ground floor of a brutalist-style civic building in the capitol complex, which is surrounded by pumping oil derricks. The standards hadn't received any press; the reporter from the *Oklahoman* had shown up to report on a high school that had violated its accreditation. But a half-dozen members of the standards writing committee sat in chairs lined up on one edge of the room. Among them was Julie Angle, the OSU professor, who participated in the 2020 standards committee as she had in the 2014 one. She'd told me a few days earlier she was not confident the revisions would make it into classrooms. "My fear is that some legislator is going to search for the word 'evolution,' find it in there, and automatically say no," she said. "But kids need to see the word evolution. They need to hear their teacher talking about evolution. It's sad that we have to tiptoe around important concepts in the interest of political congeniality." The state's curriculum director, Tiffany Neill, presented the standards to the board, focusing on the standards' structure and the work that had gone into updating them. She led the board through an experiment involving a penny, a cup, and a playing card, demonstrating how the standards focused on experimentation rather than memorization. A few of the board members asked questions, but none brought up climate change or evolution. The final vote was five yeas and one abstention.

The man who abstained was Brian Bobek, a new appointee to the board, as well as a lifelong Oklahoman, an active member of his Pentecostal church, and a longtime employee of BP Lubricants, the energy company's engine oil arm. I approached him after the meeting and asked why he had abstained. "I could not in good conscience vote yes or no," he said, then walked away before I could follow up. I emailed him later that day to ask what would have niggled his conscience had he voted. He responded, "The below from my employer should help you understand my abstention. It was appropriate in order to respect my employer and avoid any semblance of conflict of interest." Attached was BP's policies for employees who serve public office, stating that if an issue arises that "involves a matter in which BP or any of its subsidiaries has a special interest," the officer "should refrain from taking part in the consideration of these matters and be recorded as not voting on or otherwise participating in them." If having a connection to a fossil fuel company was enough to give officials pause before approving the new standards, the proposal was sure to be met with plenty of conflicting interests when it reached the state legislature. Oklahoma's largest industry, and its politicians' largest source of funding, is the fossil fuel industry.

We never found out. Nine days after the standards were submitted to the Oklahoma state legislature, the World Health Organization declared the novel coronavirus outbreak a pandemic. The legislature recessed in mid-March and didn't meet again for the better part of two months. When it did reconvene, it focused on passing a state budget. A plethora of issues that in February had seemed destined for heated public debate—criminal justice reform, medical billing reform, the funneling of tax dollars into private schools—were tabled. Certainly there was no appetite for debate on state science standards, so they quietly passed into law.

Which is to say, if Izerman's family moved to Oklahoma today, he might learn something about climate change after all.

The Textbooks

A middle-schooler who thumbed through the pages of *iScience*—a textbook that as of 2018 sat on the shelves of a quarter of American middle-school science classrooms—could be forgiven for closing the book with some big questions about recent climate change.

At the start of an eight-page section devoted to the subject, she would read that, "Average temperatures on Earth have been increasing for the past hundred years." That would be one of the textbook's only definitive statements on the subject. What has caused the warming? "Although many scientists agree with" the UN's conclusion that human industry has caused it, "some scientists propose that global warming is due to natural climate cycles." Is it a good thing or a bad thing? "A changing climate can present serious problems for society," the book states, listing several. "However, climate change can also benefit society. Warmer temperatures can mean longer growing seasons. Farmers can grow crops in areas that were previously too cold."

In some places in the sixth-, seventh-, and eighth-grade textbook series, recent climate change seems conspicuously absent. A chapter titled "Earth's Atmosphere" makes no mention of the recent state of Earth's atmosphere. A table on the pros and cons of renewable energy devotes a third of the words to pros that it does to cons, and fails to mention their role in reducing carbon emissions among them. One lesson on coral reefs and another on polar bears each explain that they face grave danger, but omit the cause of the threat. A lesson about Hurricane Katrina titled "Is there a link between hurricanes and global warming?" looks promising, but then seemingly discourages students from wrestling with the query: "Perhaps the better question is not what caused Hurricane Katrina, but how we can prepare for equal-strength or more destructive hurricanes in the future."

When the books do bring up recent climate change, they tend to provide accurate information cloaked in irresolute language. A section on nonrenewable energy says carbon dioxide "*is suspected of* contributing to global climate change."* Another chapter says "*there is evidence* that present day Earth is undergoing a global-warming climate change. *Many scientists suggest* that humans have contributed to this change." *Mights* and *coulds* abound: "More carbon dioxide in the atmosphere *might cause* average global temperatures to increase. As temperatures increase, weather patterns worldwide *could change*." This wouldn't be notable if the textbooks described other phenomena with the same qualified language, but they don't. "Volcanic eruptions *affect* climate when volcanic ash in the

* All italics in this chapter are mine, for emphasis.

atmosphere blocks sunlight." The 1991 Mount Pinatubo eruption "*caused* temperatures to decrease by almost one degree C." No "there is evidence" or "many scientists suggest" here, though scientists use the same theories and data collection devices to understand both natural and human-caused climate change.

For better or worse, textbooks exert considerable power over public education in America. We grow up believing textbooks to be an authoritative source. Many teachers look to them for guidance on what to teach and in what order, particularly when they don't feel confident in their own knowledge of a subject. Even as teachers increasingly turn to online materials, textbooks still reign over classroom learning. Per a 2018 survey, teachers in just 9 percent of high-school science classes claim never to use a textbook.

There's a lot of money to be made producing textbooks for the other 91 percent. Industry tracker Simba Information valued the 2018 market for science textbooks and other K–12 classroom materials at $885 million. The market is made up of dozens of publishers, but three dominate: McGraw Hill Education, based in New York; Houghton Mifflin Harcourt, based in Boston; and Pearson K12 Learning (now Savvas Learning Company), based in New Jersey. Together, their products made up 79 percent of the textbooks in middle-school science classrooms in 2018.

iScience was McGraw Hill's major middle-school science product of the last decade. First published in 2012 and updated repeatedly since, the three-volume set covers the basics of biology, physical science, and Earth science. Special editions have been made for Florida, Indiana, Georgia, Alabama, Mississippi, North Carolina, Oklahoma, New York, Massachusetts, and Texas. The most recent editions have a 2020 copyright. Throughout, the eight-page section with contradictory language on climate change has endured virtually verbatim, even as other parts of the books were rewritten. Who wrote those words? Why were they published? Why have they endured?

Right before John Scopes went to trial for teaching evolution in 1925, his attorney sent an urgent telegram to a New York textbook writer named Benjamin Gruenberg, asking him to come to Dayton, Tennessee, to testify as an expert witness. Gruenberg's popular biology textbook included human evolution, and the attorney thought Gruenberg might help defend the teaching of the subject. Gruenberg responded that he'd be honored to help. He then wrote to his publisher, Ginn & Company: "This is worth getting into, and it should not take more than a few days." But before Gruenberg could depart, he received a telegram from Ginn's Tennessee salesman. "I strongly urge you not to go to Dayton to testify," the salesman wrote. "It will kill your book in Tennessee and throughout the South and greatly injure Ginn and Company. I trust you will neither go nor get your name mixed up with it." Two days later, Gruenberg wrote back to Scopes's attorney. "I find that it will be impossible for me to get away for several weeks, as material from the press that I cannot shift to others is urgently awaiting my attention."

Neither Gruenberg's textbook nor the one John Scopes had used in class, *A Civic Biology*, actually did justice to the science of human evolution. To the contrary, *A Civic Biology* posited that the future of the human race depended on stopping "parasitic" families—bloodlines the book claimed were prone to illness, moral turpitude, or feeble-mindedness—from procreating. It listed Earth's "five races or varieties of man," culminating in "the highest race type of all, the Caucasians, represented by the civilized white inhabitants of Europe and America." Gruenberg's book described "white-skinned races" as "the most aggressive and masterful branch of the human family."

The eugenics and racism weren't what bothered most opponents of teaching Darwinism, but rather the theory's incongruity with the Bible. A few months after Scopes was convicted, Governor Miriam "Ma" Ferguson of Texas ordered the Texas State Textbook Commission to strip evolution from the state's new science textbooks: "I'm a Christian mother who believes Jesus Christ died to save humanity, and I am not going to let that kind of rot go into Texas textbooks." The commission demanded the removal of three chapters from one biology book. Other textbooks were told to replace the word "evolution" with "development." One member of the commission suggested removing the word "evolution" from dictionaries, but the commission decided dictionaries were technically out of its purview.

Rather than fight back, the textbook industry willingly ejected Darwinian evolution from their content (though the same could not be said of eugenics or racism). When Gruenberg's next textbook came out, it paid scant attention to evolution. Another bestseller added religious quotes to chapters that mentioned evolution. The most popular biology textbook of the 1930s asserted that Darwin's theory "is no longer generally accepted." This was no short-lived trend. Just fifteen of the fifty-three biology textbooks published between 1920 and 1959 included the word "evolution" in the text, defined it in the glossary, and listed it in the index, one study found. Generations of biology students learned from books that tiptoed around the field's organizing principle.

It was the 1957 Soviet launch of the space satellite Sputnik—the first object humankind had ever sent into Earth's orbit—that reinserted evolution into American classrooms. Fearing the US had fallen behind its rival in science and technology, President Dwight Eisenhower's administration funded the Biological Science Curriculum Study, which produced a series of textbooks that, among other things, wrote plainly about evolution. Commercial publishers followed their lead. There was more coverage of evolution in the seventeen textbooks created in the 1960s than in the sixty-six textbooks published earlier in the century combined.

Not everyone welcomed these advances. In 1964, Texas's textbook commission once again held consequential hearings on new biology textbooks. Evangelical Christians ignited their grassroots networks in protest of the offending books. The commission received 1,400 letters, calls, and petitions, overwhelmingly objecting to teaching evolution. Public debate was so spirited that the commission's meetings were broadcast on national television.

The national attention was appropriate because Texas held sway over education in other states. In the late 1800s, anger over publishers' corrupt practices and racist disputes over how textbooks discussed the Civil War had motivated Texas and other southern states to install statewide book adoptions. Previously, schools or districts chose their own textbooks. Under the new system, a state commission reviewed the books and negotiated both content and price. Texas abruptly became the largest purchaser of textbooks in the nation. Most western states also embraced statewide adoption, but northern and midwestern states never did. To this day, publishers shape their content around what will fly in Texas.

That influence was seized by a Texan couple named Mel and Norma Gabler, who in 1961 became infuriated by the contents of their son's history textbook and launched what would become a half-century-long crusade to rid textbooks not only of factual errors but of perspectives that challenged their conservative views. Norma Gabler counted among those who spoke up in the 1964 Texas hearings. "Our children will be taught atheism out of these books," she said. The commission deliberated for more than a day and ultimately approved the books. But the Gablers and their allies kept lobbying and slowly made gains. In 1970, Texas required science textbooks to include a statement declaring evolution "a theory rather than a fact." In 1974, it mandated that evolution be taught as "only one of several explanations for the origins of humankind."

Once again, textbook publishers refashioned their wares to accommodate these concerns, worried that to do otherwise might—as the Tennessee salesman warned Gruenberg in 1925—kill their books in the South. A count of textbooks' words devoted to evolution found the gains made in the 1960s were mostly lost in the 1970s and 1980s. Only in the 1990s—after states began adopting science standards, federal courts struck down "balanced treatment" acts, and the National Center for Science Education launched a coordinated campaign—did textbook publishers finally begin to treat evolution robustly and unapologetically.

Just as that happened, new concerns arose—this time, over the science of climate change. What did textbook publishers do? They produced the likes of *iScience*.

How was the distorted language on climate change written into *iScience*? Getting to the bottom of this question was no small task, since the books' title pages enumerated thirty-two authors, four consulting authors, twenty-five series consultants, twenty-six series reviewers, and twelve teacher advisors—a list that did not include in-house editors at McGraw Hill. The only thing to do was to start making calls. As I did, I pieced together that though *iScience* was first published in 2012, people at McGraw Hill began work on it as early as 2007. They reviewed the latest state standards and gathered feedback from sales reps and teachers about what should change from their old content. Then they wrote a thorough outline for the books, and contracted freelance writers to fill those outlines with content.

One author I reached, Lisa Gardiner, seemed a likely candidate to have written a section on recent climate change, since she has spent the better part of two decades developing climate educational materials for Colorado's University Corporation for Atmospheric Research. Sure enough, Gardiner told me she had been hired to write about weather and climate. However, she said, after she started, her editors specifically instructed her to *avoid* climate and focus exclusively on weather, which had disappointed her. I sent her the eight-page section to see what she thought of it. She called back, aghast. "It is just so out of date!" she said. "This is at least 1990s-level of uncertainty."

Gardiner found cause for criticism starting with the section's first paragraph, which meanders through the temperature fluctuations of the twentieth century, describing 1945 to 1975 as a "cooling period." Emphasizing small, explainable fluctuations in global temperatures is a common tactic used by denialists to muddy the larger upward trend. Gardiner also took exception to how the section devotes a full paragraph to "natural sources" of carbon dioxide—though no evidence exists that natural sources have contributed a whit to recent global warming. Then comes a sizable section on aerosols, a byproduct of burning fossil fuels, which the textbook repeatedly emphasizes can cool the climate. But analyses dating to the 1970s have concluded that fossil fuels' carbon emissions will warm the earth far more than their aerosols could ever cool it. Gardiner worried a reader of *iScience* might conclude that "we should just burn more fossil fuels and it will all even itself out."

Gardiner, like me, wondered whether the chapter's language came from its author or its editor. Regardless, she said, "clearly someone has gotten their hands on this." She wondered why the series reviewers—scientists hired to fact-check chapters in the textbook before they are released—hadn't flagged it. "I just keep thinking, who should have caught this?"

I called the series reviewers. Among them was David Ho, a professor of oceanography at the University of Hawai'i at Mānoa. In 2006, he received an email from an editor at McGraw Hill offering him $250 a chapter to fact-check their books. Curious what middle-school kids were learning, he accepted, and periodically reviewed chapters for McGraw Hill over the next few years, including some chapters of *iScience*. At my request, Ho dug through his old files. He had not been assigned to review the eight-page section on recent climate change, so could not offer insider insight there, but one chapter he had reviewed included three paragraphs that I had flagged as having the books' clearest language on recent climate change. Ho's files showed that the original draft had been more equivocal before he suggested several changes that improved it.

But in other chapters, his suggestions were ignored. One had given equal space to natural and human causes of pollution. "It's important to point out that these natural events are more rare than human impact," Ho had written. "If they're given equal billing, it diminishes the impact of humans." Nearly every other edit Ho suggested in that chapter was adopted, but when it came to climate, the original language stood. Another chapter's draft language read: "Scientists *hypothesize* that human activities are contributing to a rapid rise in atmospheric CO2." Ho had commented, "This is not a hypothesis. This is a fact." The published version read: "Scientists hypothesize that this rise in CO2 has contributed to the recent rise in global temperature." Ho told me the change did not address his concerns. "Maybe back in the early twentieth century this was a hypothesis. One hundred years later, not so much."

I sent Ho the textbook's eight-page climate-change section, and, like Gardiner, he was troubled by the uncertainty it conveyed. "What's disturbing about this is you have all these little kids who are educated this way," Ho said. "Hopefully these things can be corrected in high school or college. But why mislead them so early on?"

iScience's treatment of climate change isn't special. Competitor Houghton Mifflin Harcourt's middle-school science textbooks for Texas and Florida, copyrights 2015 and 2019, falsely describe climate change as "one of the most debated issues" in modern science. "Many people interpret" the increase in carbon to mean humans are causing global warming, the books state. Competitor Pearson's 2015 middle-school science textbook for

Texas fails to clearly define recent climate change a single time in more than a thousand pages of text.

Pearson's high-school Earth-science textbooks do something even worse. The book comes in two versions by the same authors, one for a general-education Earth-science class and the second for an advanced class. The gen-ed version devotes two pages to recent climate change. The advanced version devotes ten pages to it. The gen-ed version includes a problematic sidebar titled "Global climate change: What is causing it?" instructing students to formulate their own opinion based on internet articles with different points of view. The advanced version includes no such sidebar. This bifurcated approach has a bifurcated effect: Advanced students are thoroughly educated about the phenomenon shaping their world, while students enrolled in "rocks for jocks" are asked to decide for themselves if it's even happening.

Only a handful of academics have examined textbooks' treatment of climate change, but they have all found the same trends: The books introduce the phenomenon, thereby ticking the boxes of those who want it taught, but frame it in uncertain language, thereby satisfying deniers. A 2015 study analyzing four California middle-school science textbooks found they all introduced readers to the science of global warming, and simultaneously to the idea that it's reasonable to be doubtful of it. A 2018 study found such pussyfooting not just in science textbooks, but in social studies textbooks. Among other things, it called out how the books tend to describe global warming as a faraway phenomenon, rather than something related to the students' own communities.

I reread *iScience* with that in mind and realized that the series devotes far more words to the impact of deforestation than to fossil fuels, though deforestation contributes roughly 10 percent of annual carbon dioxide emissions, while fossil fuels contribute 65 percent. These words tend to come with photos of cleared forests somewhere in the global South. For example, a page devoted to "Deforestation and Carbon Dioxide" includes the photo caption, "These cattle are grazing on land that was once part of a forest in Brazil." No similar photo depicts the former forestlands of the United States. The eight-page section describes how the rising populations of some countries in Africa and Asia will contribute to global warming. It doesn't say that Americans have contributed more carbon to the atmosphere than citizens of any other nation, despite our comparatively modest population.

Individually, these details may seem trivial, but there's no question they shade children's understanding. "All of us have had teachers who followed the textbook lock, stock, and barrel," said Brett Levy, coauthor of the 2018 textbook study. Diego Román, coauthor of the 2015 study, noted that novice teachers lean more heavily on textbooks, and schools in vulnerable communities have the highest proportion of novice teachers. "It's one of those vicious cycles that might impact lower socioeconomic communities more, because in a wealthier district you likely have a veteran teacher who can supplement what's in the textbook with his or her own knowledge."

How much of an impact does all this add up to? Researcher K. C. Busch, Roman's coauthor on the 2015 study, conducted follow-up research with 453 middle- and high-school students in the San Francisco Bay Area. Half the students read a 250-word excerpt of a textbook by Prentice Hall (a subsidiary of Pearson). The excerpt discusses the "Climate Variation Hypothesis," which blames recent warming on solar fluctuations despite nary a flicker of evidence. Under the subject heading "Possible Effects," it asserts global warming "could have some positive effects" and other "*less positive*" effects. The rest of the students read an altered version of the excerpt in which lines like "Not all scientists agree about the causes of global warming," were replaced with "97 percent of climate scientists agree about the causes of global warming," and "less positive" was changed to plain "negative." Busch tested the students' certainty about human-caused climate change before and after reading their assigned text. They started out with roughly the same level of certainty, but afterward, those who read the skeptical text were 15 percent less likely to believe that climate change was happening. This effect held no matter how sophisticated their prior knowledge of climate change. Just 250 words sufficed to nudge the children away from clarity.

In 2020, the *New York Times* published "Two States. Eight Textbooks. Two American Stories," an investigation in which reporter Dana Goldstein dissected Texas and California editions of the same social studies textbooks. The books had the same authors and mostly the same content, but pointedly diverged in some places. For instance, the California version of a McGraw Hill history text included an excerpt from "How the García Girls Lost Their Accents," a novel about a Dominican-American family, while in the same place the Texas version featured the perspective of a border control agent.

Because the federal government leaves it up to states to run schools, publishers effectively must contend with fifty powerful ministries of education, and the changes publishers make to cater to them can have long-lasting effects. In her 1988 book,

A Conspiracy of Good Intentions, Harriet Tyson-Bernstein provided an example from 1966 in which a coffin maker and history buff was appointed chairman of North Carolina's board of education. After reviewing a history textbook for adoption, he complained that it failed to mention a minor Revolutionary War battle that happened near his hometown. By the next year, the book featured a full account of that battle. More than twenty years later, the book still devoted more words devoted to the Battle of Moore's Creek Bridge than to the Boston Tea Party or the First Continental Congress.

One evening in November 2013, the Texas Board of Education held a final meeting before approving a slate of textbooks. The last person to testify spoke against an environmental science textbook, claiming without documentation that it contained factual errors about pollution and carbon emissions. A panel had reviewed the book and had found no errors, but some board members sided with the speaker nonetheless. Who was this speaker? A geologist with the oil and gas industry and Republican candidate for the state commission that regulates that industry.

After a lengthy debate, the board voted to adopt the textbook over the geologist's objections. That outcome was, unfortunately, remarkable, said Dan Quinn, senior communications strategist with Texas Freedom Network, a progressive nonprofit that monitors textbook adoption. In some years, ultraconservatives have dominated the board and such votes have gone in the opposite direction.

Even though the book squeaked through, it's reasonable to assume the episode had a chilling effect on what publishers were willing to say about climate change in their next books, Quinn said. Before taking his current job, Quinn was a social studies editor for the publisher Holt, Rinehart and Winston, now a subsidiary of Houghton Mifflin Harcourt. While publishers want their textbooks to be factually accurate, they also want to avoid any controversy that could hurt their chances of a profit. "One editor told me one time, 'If you don't have to go there, don't. If the standards don't require you to talk about gun control, you don't talk about gun control,'" Quinn said.

This was the exact tension playing out at McGraw Hill during the development of *iScience*. One of the people who answered my calls was Adria Carey Perez, who in the mid-2000s worked as an associate textbook editor for McGraw Hill. Their earlier middle-school science textbooks had not discussed climate change in any detail. But, in 2006, Al Gore's film, *An Inconvenient Truth*, introduced the climate crisis to a broad audience, and in 2007, the UN IPCC predicted a 1.8- to 4.0-degree Celsius rise in global temperatures by the century's end. The textbook team aimed to "wade into those waters in a way that was truthful and meaningful," she said, but they were dogged by a question: Would Texas buy a book that teaches the science of climate change? "I never heard anyone explicitly say, 'We can't talk about environmentalism because of Texas.' But we all kind of knew," she said. "We had quite a few training sessions where they said, 'These are the facts we know about climate change. We can say these very few things definitively in our texts. And then there's all these things that are questionable and we have to be very careful in how we say them. In some places we won't say them at all.'"

Eventually my phone calls solved the mystery of the eight-page section on recent climate change. Its author was Jonathan Kahl, an atmospheric science professor at the University of Wisconsin, Milwaukee—just across town from the office at Marquette University where I was spending a year on fellowship. He told me that in October 2007 he had been approached by a McGraw Hill editor asking if he'd like to write some chapters for their new books. He agreed, and was sent an outline of what each chapter should include. I was eager to hear more, and he invited me and a student journalist I was working with to visit his office for an interview. He also generously dug up and shared his draft chapters, which included notes from his McGraw Hill editor.

I had been so curious whether the fuzzy language was written by its author or by an editor trying to pass Texas's scrutiny. As I read through Kahl's drafts, it became clear the answer was both. For instance, Kahl had written, "The IPCC's conclusion is somewhat controversial," referring to the UN science panel's consensus on a human cause. "Although most scientists agree with it, many believe that global warming is due to natural climate cycles." His editor seemed to suggest even that wording may be too strident. "Jon, I'm unsure how strong a statement we will be allowed to make here. I really love how you have handled the issue here, and I'm not suggesting any changes. However, this could be watered down greatly by the powers that be due to sales concerns," read the editor's comment. "Okay," responded Kahl. In the end, the language largely stood, with some tweaks. The "somewhat controversial" line was axed. "Although *most scientists* agree" with the UN, "*many believe* . . ." was altered to "*many scientists*" and "*some believe*."

A student and I met Kahl in his office. A collection of beer bottles he had collected while attending scientific conventions lined the windowsill—classic Milwaukee decor. Children's drawings of storm clouds were tacked to the wall. Kahl wore a colorful dragonfly shirt. Unruly eyebrows framed his glasses. I asked whether he would have included global warming in the textbook at all if it were up to him. Yes, he said. "And I would have included it in what I thought would be a noncontroversial way, you know, saying that anthropogenic climate change certainly happens, and to what extent the climate changes observed in the world are anthropogenic hasn't really been nailed down yet—at least [it hadn't] as of ten years ago," he said, contradicting the UN's conclusion he had written about.

Has it now been nailed down? I asked. The evidence "has tilted over towards people having a noticeable effect on climate" but "that doesn't mean that other natural cycles aren't also occurring superimposed on that in a complementary way or in an opposite way," he said. "So should climate change be included in a textbook? Yes, as long as we describe exactly what climate change is. Climate change is natural, it's normal, it's always changing, and there are different cycles and different causes." He agreed with the consensus that humans are enhancing the already existing greenhouse effect, but felt strongly that children must also learn about the natural greenhouse effect and the nonanthropogenic systems that make the climate variable. "It's very difficult to convey these things in a fair sense, especially in a textbook for kids," he said. I asked him what he would consider unfair to convey. "[That] the greenhouse effect is due to anthropogenic emissions and that is causing temperatures to rise and so we should stop global warming. That would be unfair. Because the greenhouse effect is a natural phenomenon," he said. "And also the notion that climate change is bad.

Climate change has always occurred and always will occur. And I suppose, depending on your perspective, certain changes could be good, certain changes could be bad." I asked him to clarify—was he saying that textbooks shouldn't include any judgments about whether human-caused climate change is good or bad? "Well, certainly I don't think that politics has any place at all in a science textbook," he said.

In some ways, Kahl was an obvious choice to write a chapter on climate. He had participated in a host of science education projects and authored several books for children. Since the mid-1980s, he has been conducting research on atmospheric trends, including some work sponsored by the National Science Foundation and the National Oceanographic and Atmospheric Administration. But a look at his CV revealed that three of his earliest grants came from the Electric Power Research Institute, a think tank with close ties to the infamous climate change denial industry group, Global Climate Coalition. Between 1985 and 1993, Kahl received $195,000 in research grants from the group. At least one of the resulting studies has often been cited by climate deniers.

I asked Kahl about this connection. He told me he had applied for Electric Power Research Institute grants "because I had some sense that they would be interested in funding the research I wanted to do," and that the organization had exerted no editorial influence over his research, in accordance with the terms of the grants. Regarding the study often cited by climate deniers, "that was a very strange event, the politicization of that article," he said. His research had looked at a forty-year dataset of temperature readings over the Arctic Ocean and had found no evidence of warming. After the research was published in

1993 in the journal *Nature*, he made the point in media interviews that the absence of evidence did not equate to evidence of absence. Nonetheless, he said, the research was oversimplified by mainstream media and seized upon by Rush Limbaugh and others in the right-wing media to push a narrative that climate science was suspect.

He told me it was "preposterous" to suggest that what he wrote for *iScience* somehow supported the views of climate change deniers: "I just have to roll my eyes at that." He said he considers himself an environmentalist, he fully accepts the evidence that humans are the primary cause of recent climate change, and he believes humans must take action to protect the environment. He ascribed the qualified language in the chapter to his tendency to see systems as complex. "As a meteorologist, I know that things are never as simple as, 'Oh, this is anthropogenic, or it's not.' It might be mostly anthropogenic but there's always natural cycles," he said. "Sometimes I have trouble simplifying things to a level that's consistent with how other people think."

The last person I chased down was the editor who had made the notes on Kahl's chapter. Gabe Langbauer worked at McGraw Hill from 2006 to 2008 as an assistant editor. He had recently graduated with a master's in meteorology, and was excited to participate in what he imagined would be a sophisticated process. "I had a vision before I took the job of textbooks being written by a huge team of PhDs, who studied every single word and made sure it included the best and the latest science. And then I showed up with this meteorology degree, and they say, 'Okay, edit this chapter on RNA.' And I say, 'Oh, okay. I guess I can google things.'"

One day, his boss handed him a sheet of about twenty names of potential authors for the weather and climate chapters—all people who had publishing experience in the field. He spent a few hours doing a quick internet search on each one and narrowed the list to a few names. Jonathan Kahl's experience writing children's books made him stand out. Langbauer told me that if he had thought Kahl might have even loose connections to denialists, he would never have hired him.

I read Langbauer some excerpts from the eight-page section on recent climate change he had edited. Even over the phone, I could sense him wincing at the line, "Although *many* scientists agree" with the IPCC, "*some* scientists propose" recent global warming is natural. "That is not a good statement," he said. When it was written, there might have been a small number of legitimate scientists still questioning the human causes of global warming, but if so, it was only 1 or 2 percent, he said. "I would say that statement is factually accurate but intellectually dishonest. It's a defensible statement that would stand up to a court of law, but you're not getting to the heart of the matter." Like Perez, Langbauer recalled explicit conversations on how to write about climate change, mainly about "how to be factually accurate and not piss somebody off." It wasn't that people had "some secret agenda to sneak in the oil industry perspective," he said. "It was a matter of practicality, like, 'If we're aggressive about this stuff, Texas isn't going to buy this book.'"

I read him the editor's notes on Kahl's draft, including the one about the "powers that be" watering things down. "Yeah, that sounds exactly like what I would have said," he said. "I don't remember specifically, but I am assuming that my boss or someone else in the editorial department had read it and said,

'This is more than we can say.' So in that comment I'm saying, '*I'm* not going to change it, but just so you know, it might get changed.'"

I asked him how he felt about the final product. He said he couldn't know if publishing weak language about the climate crisis was the right business decision. "If we said that it was extremely likely that human-caused climate change is real, what's that do? Does McGraw have enough clout to just sell the book anyway? Or then do those states just go with Prentice Hall and no one buys McGraw Hill?" But from a scientific standpoint, it was obviously wrong, he said. "I really wish we had been more aggressive in our wording."

I sent McGraw Hill's communications department a long list of questions. "As part of our commitment to academic integrity, we strive to ensure all of our instructional materials are grounded in the most current research and guidance available to us at the time of publication," they responded. "One of the challenges of writing science curriculum is that scientists are trained to be skeptics and to check and re-check, and are frequently wary of making definitive statements of fact. When we capture science in real time, there is likely to be some evidence that comes back after the fact that refutes science. That's where K–12 education was on climate change ten to fifteen years ago. The IPCC was still in its infancy. We have plenty of evidence now and did then about the human impact on climate change, but there were some scientists that were skeptical. And we thought at the time that we should take a moderate approach to let teachers choose how to present the material. Our approach with *iScience*, which was written between 2007 and 2009, was to take a cautious approach, allowing teachers and students to have discussions about the topic. The 2007 IPCC report, upon which most of this *iScience* content on climate change was derived, was not as definitive as the later 2014 report."

In fact, the IPCC was well into adulthood by the time the textbook was written, having been established in 1988. The 2007 report said the evidence was "sufficient to conclude with high confidence" that humans were changing the climate, while the 2014 report said that humans "are extremely likely to have been the dominant cause" of the warming. Regardless, McGraw Hill did not update the skeptical wording even after the 2014 report, and as of 2021, still markets and sells *iScience* to districts around the country.

McGraw Hill is also marketing and selling a second set of books. In 2019, the publisher announced a new K–12 science series called *Inspire Science*, designed to align with the Next Generation Science Standards. Sophisticated climate educators could find fault with its treatment of the subject. Most of its middle-school climate coverage shows up in a single lesson at the end of a module in sixth grade, so if a student misses a day or two of school, or if a teacher chooses to skip the section, the student may not learn about it at all. Like *iScience*, *Inspire Science* emphasizes deforestation over fossil fuels, and at one point asserts without evidence that "the amount of carbon dioxide in the air has increased due to natural and human activities." (In response to these criticisms, McGraw Hill wrote that the series was designed to meet the requirements of the NGSS, "and as such, coverage of climate change is placed in the appropriate points in the learning progression. Theoretically, we could weave climate change through other parts of the program—however, this isn't something that has been requested by our middle-school customers.") Those criticisms notwithstanding, *Inspire* certainly treats the climate crisis far more thoroughly, directly, and accurately than its predecessor. The lesson on the climate crisis has been expanded from eight pages to twenty-six pages. The unit provides charts of temperatures and carbon dioxide levels over time and asks students to deduce the relationship for themselves—a useful learning tool. None of Kahl's language remains.

Now, a divide is opening. Publishers like McGraw Hill create and market two separate products: One, textbooks like *Inspire Science*; the other, "legacy" textbooks like *iScience*. The loose red-blue partisan divide between states whose standards accept climate science and those whose standards avoid it informs the textbooks they adopt. Ideology trickles from elected officials through education departments and into classrooms. Accurate information about climate change thus becomes the purview of children living in liberal states, while children living in conservative states are frequently provided fodder for denial.

Selling Kids on Fossil Fuels

One early spring day I was sitting in the science department of an Arkansas middle school when a representative of the state's oil and gas industry walked in. She was there to talk to the seventh graders.

Her name was Paige Miller, a petite blonde with a short shag-cut and big silver jewelry. She told me she ran Arkansas Energy Rocks!, an initiative of the Arkansas Independent Producers & Royalty Owners, which describes itself as "the voice of Arkansas's oil and natural gas community." Twenty attentive tween faces watched as she cued up a PowerPoint presentation.

Arkansas, she told the students, had the good fortune of being an energy state. She described the layer cake of earth and minerals under the students' feet and pointed to diagrams showing which of those layers are soaked in fuel. She showed them pictures of the technology that sucks that fuel out. Twenty-five of the state's seventy-five counties produce either oil or gas, and some 33,000 people are employed in that production. Arkansas is a landlocked state, but some Arkansans even work on offshore oil rigs, she said. They fly to the rigs on helicopters, work for two weeks, then get two weeks off.

"Does that seem like a good schedule? Half time?" she asked the students.

"Mm-hmm!" the kids chorused.

"So how much do they pay you?" asked one student.

"The average starting salary on an offshore drilling rig is $100,000 a year," Miller said.

"Wow!" whispered another student.

"Fossil fuels have been very important to mankind," she said, and launched into a list of ways that that is true. She showed the students a pie chart of the nation's energy sources. Fossil fuels made up the majority of the pie. Each renewable energy source comprised a slim slice. But using fossil fuels comes at a cost, she told the students. "The problem with fossil fuels is carbon emissions," she said, without elaborating on the nature of that problem. "But somebody's going to have a problem with all of these energy sources," she said. "Geothermal power works well but it's expensive. Wind power a lot of people don't like because they say it kills birds. A lot of people don't like hydropower because they say we shouldn't be damming up bodies of water. With solar, if there's a tornado, what happens to the solar fuel?"

"It goes away," the students said.

"You're going to find a problem with any one of these sources."

As for the carbon problem, there's not much the US can do about it, she said. "If the United States shut down all fossil fuel usage tomorrow—all of it—the difference it would make in terms of global warming is 0.01 percent," she said, inaccurately. She did not define global warming, but then presented a dark scenario of what might happen if we address it. "There are actually 1.2 billion people on the continent of Africa and the country of India who live every day without electricity. Now think about that—what your life would be like without electricity. You don't have a refrigerator in your house full of good food being cooled properly. If you have a bad accident, you may not be able to get to a hospital that has electricity in time to save your life. Access to energy is literally the difference between a first-world country like the United States and a third-world country like India. The problem is they didn't build out their infrastructure like we have."

So, when you consider energy, you have some real thinking to do, she told the students. "First of all, you need to decide your standard of value. You need to decide: Is human life the most important? Humans getting healthier, wealthier, happier, living longer? Or is pristine nature more important? Do we want to quit building new houses? Stop getting stuff out of the ground? Do we want to leave it exactly as it is? Because that would be difficult. Thankfully, we don't have to choose in this country. We're working on a happy medium at this point."

The students didn't ask any questions, and she moved on. "While there are challenges with fossil fuels and we have to solve those problems, they have provided us with the ability to make lives better all over the world," she said. As she concluded, she gave out pencils printed with "arkansasenergyrocks.com," and encouraged everyone to visit the website.

For a portion of her career, Miller gave presentations mainly to adults. When fracking started in Arkansas, people had a lot of questions about its implications, so she would visit Rotary Clubs and city councils to provide the industry's answers. But these days, she gives her presentation to any classroom that will have her. Mostly, she says, she speaks at middle schools. Frequently she gets third or fifth graders. Sometimes she gets the odd high-school class. It's a lot of time on the road, but it's worth it, she said. "We decided, you know, if we *really* want to change the way people look at energy sources and understand what we use and why we use it, we need to go out and work with teachers and talk to students."

The phenomenon of fossil fuel companies plying schoolchildren with their messages is decades old. The American Petroleum Institute was making the case for marketing to children as early as the 1940s, according to archives reviewed by the Center for Public Integrity. A survey of 10,000 Americans had indicated the industry's reputation could use some rehabilitation, and a "well-directed program of public education" could help. To that end, API teamed up with DuPont and by 1954 had trained 600 oil industry workers to give a show-and-tell program called "The Magic Barrel" to schoolchildren. In 1972, General Motors published a booklet to counteract what its pollsters said were children's "negative" attitudes toward auto companies. The booklet featured cartoon characters "Charlie Carbon Monoxide" and "Harry Hydrocarbon" (a "harmless demon") who helped dispel fears that air pollution could lead to serious health hazards. By the next June, the company had distributed 2.1 million copies of the booklet, including to 62,000 elementary-school principals.

In the midst of the 1970s oil crisis, Exxon's public affairs department partnered with Walt Disney Educational Media Company on comic books about energy conservation. In one,

Mickey Mouse and Goofy Explore Energy, the pair get in trouble when their car runs out of gas on a fishing trip. On their walk to the service station, they learn about supply and demand from a smiling nuclear symbol called "Enny, the spirit of energy!" Another comic book was included as an insert in a 1978 issue of the National Education Association's journal, which reached a million teachers.

Much more strident commentary on the crisis came from the petroleum company Amoco (later merged with BP). It produced a twenty-six-minute film titled *The Kingdom of Mocha*, ostensibly to introduce students to economic concepts. In it, the primitive "Mochans," led by a chief called "Big Daddy," become dependent on an energy source—wood. When a war cuts off trade, Big Daddy responds by threatening to impose price controls or even to take over the wood industry, which viewers learn could prove calamitous. The parable—in addition to being flagrantly racist—executed a trenchant attack on 1970s-era US energy policy. Amoco claimed that more than 20 million schoolchildren watched it.

Today, fossil fuel–funded educational programs aimed at children are abundant. A non-exhaustive search found such programs in Alaska, Arizona, California, Colorado, Florida, Illinois, Kansas, Kentucky, Michigan, Montana, Nevada, New Mexico, New York, North Carolina, Ohio, Pennsylvania, South Carolina, Texas, Utah, Virginia, West Virginia, Wisconsin, and, of course, Arkansas. Not all have Paige Millers traveling classroom to classroom. More common are free curricula, sponsored activities, and scholarships. Some promote safety: The Missouri-based utility Evergy, for instance, created an online game to teach students to recognize electrical dangers.

Others are transparent marketing efforts. To list just a few: The Offshore Energy Center in Texas offers summer camps and field trips to schools and clubs "to promote the importance (past, present, and future) of the offshore energy industry and its contributions to our quality of life." The American Coal Ash Association Educational Foundation provides scholarships for students, including one for those who "have an interest in advancing the beneficial use of coal combustion products." The Independent Petroleum Association of America has sponsored "High School Petroleum Academies" in five schools in Texas, where students pretend to be oil executives and give presentations on "misconceptions" about the oil industry. The American Coal Foundation partnered with Scholastic to create a fourth-grade curriculum called "United States of Energy," which made no mention of coal's impact on the environment or health. Utility consultant Culver Company has created activity booklets with titles like *Natural Gas: Your Invisible Friend*, filled with puzzles and games. In one, the student solves math problems to crack a code and complete the sentence, "Using natural gas to fuel vehicles is a good way to reduce . . ." with "harmful air pollutants." Power utilities pay Culver to brand the booklets with their name, as Massachusetts utility Eversource did before distributing them to schools in the Boston area.

Industry education programs in Kansas, Ohio, Illinois, and Oklahoma are actually supported and sanctioned by those states' governments. The most sophisticated is the Oklahoma Energy Resources Board, a "privatized state agency" voluntarily funded by oil and gas companies. The OERB has produced a series of videos by "Professor Leo," a goofy Bill Nye knockoff who educates students about the state's oil and gas resources.

Teachers can ask for "Petro Pros" to come speak to their classes, or tap into a library of glossy lesson plans ready-made for any age or subject. Classes that complete an OERB curriculum are treated to field trips to certain museums. A field trip to a children's museum in Seminole features a new two-story exhibit: "The lower level is dedicated to what one might find underground to discover oil. The upper level includes a refinery, Christmas tree, and a working mini-pumpjack."

For the K–2 crowd, the agency sends copies of children's books featuring characters like "Freddie Fuelless," "Oliver Oilpatch," and "Petro Pete" to elementary schools across the state. In one, *Petro Pete's Big Bad Dream*, the titular character drifts off to sleep wondering what the world might be like without petroleum products. He awakes to find his school clothes, toothbrush, and bike tires missing. He blames his dog and heads to school in his pajamas. At lunch, ice cream spills out of the soft-serve machine as liquid and there's not a soccer ball to be found. Finally, his teacher figures out what has happened: "It sounds like you are missing all of your petroleum by-products today!" she says. "Having no petroleum is like a nightmare!" Petro Pete declares, before waking up and realizing it was all just a dream.

Another book features Boomer Burrow, home to a menagerie of wild animals. One morning, bulldozers interrupt their peaceful burrow. But Petro Pete's dog RePete explains that they are there to clean up old oil equipment. The animals are skeptical: "Do you mean to tell us these humans are cleaning up old oilfield messes just because they want to?" asks a rabbit. "Yes! The humans are taking responsibility for their industry and cleaning up abandoned well sites in Petroville and all across Oklahoma," RePete assures him. A few weeks later, the animals celebrate their newly clean neighborhood. "Let's hear three cheers for RePete and the humans!" cries the rabbit.

The OERB has spent roughly $50 million on K–12 education since 1996 and reached an estimated 3.3 million students. Ninety-eight percent of Oklahoma school districts use OERB materials. One obvious driver of this popularity is the agency's largesse: Teachers who spend a half-day in an OERB training seminar are rewarded with at least $300 worth of classroom lab equipment. In a state where funding for education has run so low that some districts cut their school week to four days, 17,000 teachers—more than a third of Oklahoma's teaching force—have participated in the training.

Some funding for energy education comes from the US Department of Energy. Their grants often go to conservation programs, like Montana's "SMART Schools" competition, which rewards schools that conserve energy. But taxpayer money has also been used to promote industry interests. In one such instance, elucidated by an *Austin American-Statesman* investigation, a federal grant seeded a program called the Energy Education Project. The project was the brainchild of State Rep. Jason Isaac, who had spotted a question in his child's schoolwork that asked, "Which of the following fossil fuels causes global warming: oil, gas, coal, or all of the above?" Isaac, who sits on the board of the Texas Natural Gas Foundation and whose campaigns received $43,501 in donations from the oil and gas industry between 2013 and 2017, said the question made his blood boil. "It should have been none of the above, in my opinion," he told the *Statesman*. "It's such a biased question. It's making their minds up for them. It's very negative. You're striking fear in children that oil and gas and coal are bad."

The foundation applied for and was awarded $165,000 in federal grants, administered by the state of Texas, to develop new classroom materials. The resulting curriculum told students that it remains unclear whether renewable energy is actually better for the environment than nonrenewable energy, and declares that ending our dependence on oil and gas would be "devastating to us socially as well as economically." A science education expert who reviewed the material for the *Statesman* described it as "notable for its lack of actual science content."

In years past, some energy education efforts blatantly rejected climate science. In 2002, the American Petroleum Institute launched the domain www.classroom-energy.org. That site is now defunct, but much of it is still available on Internet Archive's Wayback Machine. In addition to industry promotions—lesson plans like "Discover the Wonders of Natural Gas" and "There's a Lot of Life in a Barrel of Oil"—the site devoted a page to climate: "It is estimated that all human activity, including all combustion—for transportation, building heat, power generation, industrial manufacturing—generates less than five percent of total atmospheric carbon dioxide," it states incorrectly. (When the site was launched, humans were responsible for about 19 percent of the carbon dioxide in the atmosphere; twenty years later, we lay claim to about 28 percent of it.)

The Rocky Mountain Coal Mining Institute, whose tagline is "Promoting Western Coal Through Education," had a "Global Warming Quiz" on its website as recently as 2020 that began with the disclaimer: "Caution: This section contains sound science, not media hype, and may therefore contain material not suitable for young people trying to get a good grade in political correctness." The first question is true or false: "'Global warming' is a real phenomenon: Earth's temperature is increasing." Whatever you answer, you get this explanation: "'Global Warming' is something that has been happening for a long time. The temperature of the Earth has been increasing more or less continuously since the time of the cave man. . . . Don't panic when you hear global alarmists warning the Earth may have warmed almost 1 degree in the last 200 years. Although this still hasn't yet been proven, it is in fact exactly what should be happening if everything is normal. If Global Warming stops, then you can start worrying! It means our warm interglacial phase is over and we may be heading into another Ice Age!"

Another "quiz" with questionable answers spent several years up on the federal Bureau of Land Management's website for the Campbell Creek Science Center in Anchorage, Alaska. In an echo of Paige Miller's approach, the quiz asked children, "What's more important, (A) fulfilling our nation's vast and varied energy needs or (B) Protecting natural ecosystems and ensuring recreational opportunities?" Those who chose "B" were asked: "Hmm . . . are you sure?"

The Canada-based Fraser Institute, historically supported by the Charles Koch Foundation and ExxonMobil, in 2009 produced a booklet including six lesson plans on climate change, which said things like: "Despite higher CO_2 levels in recent years, global temperatures are now expected to remain stable or even decline on average in certain regions." Referencing a radically inaccurate chart, the lesson says, "Temperatures do not appear to have risen because of changes in atmospheric levels of CO_2. The relationship between these two variables demonstrates that correlation does not imply causation."

In more recent materials, when climate change is discussed, it tends to be ignored or mentioned in passing. For instance, the US Department of Energy's site for teachers, "Energy Literacy: Essential Principles for Energy Education," mentions the subject to say that issues like climate change "highlight the need for energy education." But after using the issue to contextualize its own relevance, the curriculum does not define or discuss it. The Rocky Mountain Coal Mining Institute's educational materials talk about clean coal but have nary a mention of climate change. Likewise, Anchorage's Campbell Creek Science Center's site features dozens of educational resources, but I couldn't find any discussing climate change. Anchorage's average winters are eight degrees Fahrenheit warmer than they were seventy years ago.

The biggest player in the energy education scene these days is the National Energy Education Development Project, a $4.7 million nonprofit established in 1980 whose mission is to "promote an energy-conscious and educated society." Its sponsors and affiliates include energy interests of every stripe, from the American Wind Energy Association to Shell. A portion of group's funding comes in the form of state and federal grants.

The organization hosts teacher trainings, runs student workshops, and maintains a large catalog of classroom materials. NEED spokesperson Emily Hawbaker told me that "renewables and energy conservation makes up the majority of what we do."

I reviewed dozens of educational materials on NEED's site to see how they handle the crisis that scientists and economists expect to transform the world's energy future, but found sparing discussion of it. A set of "Energy at a Glance" factsheets fails to mention climate change once, as do the thirty-nine science fair project ideas NEED offers. In a page called "Helpful Energy Sites," the group links to many of its affiliates—including the American Petroleum Institute—but provides no links to information about climate change.

NEED's main curricular offering is a collection of "Energy Infobooks": hundreds of pages of materials and activities introducing learners to energy. Among them are a two-page booklet on climate change for middle-school kids and another two-pager for high-school kids. None of NEED's fourteen pages of lessons on petroleum discuss the fuel's influence on the climate, or even mention carbon dioxide. Its lessons on natural gas repeatedly emphasize how nonpolluting that fuel is, though researchers have established that methane leaks in natural gas infrastructure are a major contributor to atmospheric greenhouse gases, and the UN's climate science panel has recommended cutting global use of natural gas between 13 and 62 percent before 2050. This may be lost on first graders who participate in the NEED lesson that has them chant: "Burn CLEAN [holding up one hand], burn FAST [bringing the other hand to meet the first], NATURAL GAS [moving hands upward to make the shape of a flame]!" Among the materials on fossil fuels, climate change only substantively appears in a booklet on coal, which provides a reasonable explanation of the greenhouse effect and notes that burning coal adds carbon dioxide to the atmosphere. But from there, its language can only be described as industry-friendly. "Most scientists believe the Earth is already experiencing a warming trend due to the greenhouse effect," it states, as though the warming trend hasn't been verified by millions of thermometers around the world. In a section on "cleaner coal" (a concept long promoted by the industry in an effort akin to the merchandising of "safer cigarettes") it talks up the promise of carbon capture, a much-studied technology that has never met success in practice. The nation's sole carbon capture coal plant was mothballed in 2021.

In the past, NEED's discussion of climate change was far worse. A lesson for primary and elementary students called "Energy Stories and More," published in 2014 and available on the site until 2017, said that "some scientists think it's too soon to tell if the Earth is really warming. They think a little warming might be a good thing for the Earth. What do you think?"

I asked NEED's executive director, Mary Spruill, why such language would exist in modern curricula for children. She said they've had to handle the subject delicately out of respect to educators worried about teaching the "controversial" subject. "Many of our teachers will say, 'You know what, I might get pushback if I try to teach that.' So we do our best to make sure everything is in a format that teachers are going to be able to teach in the classroom with little pushback from anywhere and really make it work for the students." In fact, she said, NEED had just developed a grant-funded curriculum with partners in Rhode Island about heat islands and health impacts of climate change. "And even in that project, we were told that some districts are still saying 'nope' to teaching curriculum like this. Others are saying, 'Yes, give it to me, let's do it.' So I think that's why you'll often find qualifiers." I pointed out that scientists would push back on the idea that one should water down science education to accommodate the politics of some adults. "I think it's a fair point you're making," she said, and added she would ask her staff to review the materials' language.

I asked Spruill her thoughts about partnering with the fossil fuel industry on events like the "Phillips 66 Virtual Energy Education Workshop," a professional development seminar that pays teachers a stipend (or reimburses their schools for a substitute teacher) and sends participants home with a $300 kit called "The Science of Energy." She prickled. Many industries get involved in education, she said. She brushed off the characterization that NEED's goal is to promote industry messages. "We don't sit in meetings and create marketing materials," she said. "We're a nonprofit. When we're able to find a sponsor willing to sponsor something, we are grateful. We could not get teachers into workshops if we didn't have someone willing to sponsor them."

When I contacted NEED several months later to fact-check this material, they again pushed back on the characterization that the organization is a mouthpiece for its sponsors or that it downplays global warming. To the contrary, wrote spokesperson Hawbaker, the team is "very passionate about putting out quality resources that engage students in becoming responsible energy consumers, and yes, taking action to tackle climate change. . . . It's probably very easy for you to assume we have a bias and engage in irresponsible efforts, simply because of a sponsor you noted on our sponsors list. However, this is simply an incorrect assumption. . . . Partners do not come to us for puff pieces, PR stunts, or marketing messages, because we will only present the science." Furthermore, she wrote, "it may be true that some partners we work with have played a part in creating the problems we wish to change as a society, but if we don't allow them to participate in making positive change, are we doing all we can to help?"

I heard Mary Spruill's reasoning again and again: Our underfunded schools need all the help they can get; public-private partnerships infuse money into education; the fossil fuel industry is not alone in making educational materials for schoolchildren. All those statements are true and have been for a long time. As documented in the 1979 book *Hucksters in the Classroom*, at the same time elementary-school principals were opening their mailboxes to find Charlie Carbon Monoxide, home economics teachers were opening theirs to the recipe book *Cooking with Dr. Pepper* and the film *Mr. Peanut's Guide to Nutrition*. More than forty years later, attend any teaching convention and you'll find yourself wandering massive halls filled with booths. From many of them, private companies distribute free educational materials. Proponents of such programs say educating students about industry not only helps out resource-hungry educators, it can make students aware of potential careers.

The trouble is that public-private partnerships don't always result in quality education. An organization of educators called the Climate Literacy and Energy Awareness Network, cofounded by the National Oceanic and Atmospheric Administration's Frank Niepold and largely funded by grants from that agency, surveyed some 30,000 lesson plans and resources about climate change available for free online. It found only 700 acceptable for use in schools. The rest were outdated or scientifically unsound. Teachers can be excused for not always discerning the good from the bad, said researcher Eric Plutzer. "You can imagine that a teacher who is overworked and asked to solve all of society's problems will be open to curricula that other people write—especially one with a nice lesson plan with visuals and suggestions for student exercises." The industry education programs can thus exploit the trusted relationship between a learner and their teacher in order to implant their messages into the minds of children.

In the class Paige Miller spoke to, the teacher was deferential to the guest and expressed no critique or counterpoints to what she was saying. Not surprisingly, the students didn't either. To be sure, Miller's presentation included accurate information about the science of fossil fuel extraction, but it ignored the science of that extraction's impact on the world. It downplayed the outsized contribution Americans make to atmospheric carbon levels. It talked about the birds killed by windmills without mentioning that as many as two-thirds of American bird species may be extinguished by climate change this century. It talked about the high salaries of offshore drillers but not of the average Tesla employee. It tendered a condescending, colonizing view of the "third world," then inaccurately presented fossil fuels as the solution to its problems (there are many reasons why some impoverished regions lack electricity, but none of those reasons involve the adoption of renewable energy). When Miller asked the children to choose between human well-being and "pristine nature," the students would be forgiven for equating fossil fuels with human well-being, when in fact unmitigated fossil fuel use will not only shatter whatever is still pristine in nature but will also exact a bitter toll on the well-being of humans.

What's more, this presentation took place in Helen Tyson Middle School in Springdale, Arkansas, home to the largest enclave of Marshallese people on the continent. As the class wrapped up, I asked the teacher to point out the Marshallese children in her class. She pointed out five, a quarter of the students present.

The Victory

A month after Izerman sat in a bright Marshallese classroom staring up at crayon-drawn posters connecting pollution, melting ice, and rising seas, scores of adults sat in a dim Washington, DC, conference room staring up at a PowerPoint presentation containing the following bullet points:

- Arctic sea ice has decreased since 1970s
- But it was increasing from the 1940s to 1970s!
- IPCC based their conclusions on poor data and computer models that didn't work
- Arctic sea ice has been repeatedly advancing and retreating since long before the Bronze Age!
- It seems to be due to the Sun, not CO2

Lest there be any confusion: Data shows only a marginal increase in sea ice between the 1940s and 1970s and the ice "gained" in that period has been dwarfed by the amount lost since then. The UN IPCC's models have been remarkably accurate. There is no evidence supporting the notion that the sun, rather than the growing amount of carbon dioxide in the atmosphere, has caused the recent extraordinary loss of ice in the Arctic. Which is to say, the crayon posters were considerably more accurate than the PowerPoint.

The DC conference was hosted by the Heartland Institute, a libertarian self-described "action tank" best known for its commitment to promoting doubt about climate change. Roughly once a year since 2008, the organization has hosted an "International Conference on Climate Change," bringing together all the loudest voices in the climate denial movement. The voices do not sing in unison. Some question whether global temperatures are really rising, accusing scientists and environmentalists of misinterpreting, exaggerating, or corrupting the data. Others grant that warming is happening, but say it's natural. A third group admits it's happening and is human-caused, but insists it will somehow *benefit* the planet. All generally agree that humans can endlessly raise concentrations of greenhouse gases in the atmosphere without a major and damaging climatic response, and that to suggest otherwise is climate "alarmism."

I slipped out of the darkened room and found Joe Bast, Heartland's then-president and CEO, in the hallway. I told him I was especially interested in what kids learn about climate change in school and described a survey published in 2016 revealing that nearly half of the science educators who teach climate change tell students it is natural, debatable, or say nothing at all about its cause. "I don't think that was the result of anything we did," Bast demurred. "Science teachers are smart. Smarter than the average reporter."

Then Bast told me that Heartland was about to mail a book called *Why Scientists Disagree About Global Warming* to every science teacher in America. He gave me a copy, as well as a DVD called *Unstoppable Solar Cycles* and a letter that would also be included in the packages. Like so much climate denial literature, the materials did their best to mimic science, featuring graphs, charts, footnotes, and even references to peer-reviewed studies. But despite those trappings, they would not have survived peer review themselves. Under peer review, a scientific work is scrutinized by a group of scholars from a relevant field. The reviewers look for problems: Data that has been fudged, misinterpreted, or cherry-picked. Assumptions that cannot be tested. False equivalencies and other logical fallacies. Experiments whose steps aren't explained well enough to understand or replicate. The materials Heartland sent to schools would have run into trouble on most of those fronts.

After I wrote about the mailings for *FRONTLINE* and the GroundTruth Project, public outrage ensued. Four Democratic US senators wrote to then-education secretary Betsy DeVos—whose family foundation has funded climate denial groups—asking whether her staff had contact with Heartland. Several ranking House Democrats issued statements demanding Heartland stop its campaign. "Lying to children about the world we live in to further corporate polluter profits is cruel," said Rep. Raúl M. Grijalva (D-AZ). But the Heartland Institute had no intention of desisting. "Is this a belated April Fools' Day joke?" Bast wrote in response to the lawmakers' statements. "If not, it should be. This is hilarious." Heartland says it eventually mailed more than 200,000 packets out. Many of the science teachers I spoke to while reporting this book had received one.

It's unclear how much the book and DVD were actually used in classrooms. A Heartland representative said they had no metrics to share about how widely the materials were used, "but we consider it a great success." In my reporting, I never met someone who had themselves used the book, though a couple said they knew a colleague who had. Plutzer's 2019 survey of 1,427 science teachers had asked them to list three educational resources they used to teach climate change aside from textbooks; the National Center for Science Education's Glenn Branch reviewed the responses and found only a few mentions of denialist literature. Frank Niepold, NOAA's longtime climate education czar, said most teachers he spoke to after the campaign had immediately spotted the materials for what they were, and some had even used them to help students identify propaganda. But he didn't doubt that the materials had found their way into some classrooms. Casey Meehan, a former teacher turned academic who has studied climate education, said the same. "When I was teaching, I barely had three minutes of my day to pee. So actually having something mailed to me that looks highly produced, from what looks like a reputable organization—I get it," he said. "Frankly, it's a brilliant ideological lobbying technique."

In the mid-twentieth century, it began to dawn on the tobacco industry that it had a major problem. Its own research showed that cigarettes were linked to a plethora of bad health outcomes, including death. The industry had some options. It could let people know of the dangers. It could at least refrain from peddling its addictive wares to children. Instead, tobacco companies chose to hide the problem for as long as possible, while continuing to profit.

We know a great deal about how this all happened because later lawsuits forced the companies to make their records public. Among the people who dug through those records were Naomi Oreskes and Erik M. Conway. What they found became the basis for their 2010 investigative book, *Merchants of Doubt*. The industry had developed a strategy to systematically deflect attention from unfavorable scientific conclusions. They hired scientists who went on TV and wrote op-eds questioning whether cigarettes were truly unhealthy. They also poured millions into funding research that could cloud the prevailing conclusions.

To execute this strategy, one needed willing scientists. *Merchants of Doubt* found that a small community of scientists were responsible for the majority of the tobacco industry's doubt campaign. A key character was solid-state physicist Frederick Seitz, who had contributed to the development of the nuclear bomb and served as president of the National Academy of Sciences. In the late 1970s, he went to work for R.J. Reynolds Tobacco Company, and helped them distribute $45 million to scientists between 1979 and 1985, for research that could be used to defend the company in court.

Other industries threatened by the conclusions of science began replicating tobacco's doubt-mongering strategy. Another physicist, S. Fred Singer, became a mainstay in the cast of scientists who aided in these efforts. Singer had helped develop Earth observation satellites in the 1950s and later worked for the National Weather Satellite Center, the federal departments of commerce, transportation, and the interior, and the Environmental Protection Agency. In 1982, Singer was appointed to a federal committee investigating the problem of acid rain and became the sole member to argue that its cause was inconclusive. Both Singer and Seitz railed against the science behind the "nuclear winter theory," which established that a major exchange of nuclear weapons could radically alter the climate. Singer questioned the science linking asbestos to health problems. When scientists discovered that a human-made chemical called chlorofluorocarbons (CFCs) was gnawing a hole in the ozone layer, Singer accused those scientists of self-interest. "It's not difficult to understand some of the motivations behind the drive to regulate CFCs out of existence," Singer wrote in a 1989 article in *National Review*. "For scientists: prestige, more grants for research, press conferences, and newspaper stories. Also the feeling that maybe they are saving the world for future generations."

Oreskes and Conway note that both Singer and Seitz stopped conducting original research in the 1970s, shifting to interviews and editorials in which they disparaged science being done in other fields. Regardless of the subject, they invariably concluded that regulation was inappropriate. These men were physicists, not medical researchers or epidemiologists, but journalists and politicians nonetheless took them seriously on these questions, fooled by their resumes and the assumption that someone smart in one field must be smart in all fields.

When the fossil fuel industry realized that global warming could threaten their business model, it knew who to turn to. Both Seitz and Singer became bright stars in the small galaxy of climate denialists, relentlessly and publicly arguing that modern climate science was wrong. Among Singer's many publications was *Why Scientists Disagree About Global Warming*—the book he coauthored that in 2017 landed in the mailboxes of science teachers across the country.

As discussed in chapter 1, in March 1998 the American Petroleum Institute hosted a meeting with the explicit goal of promoting doubt about recent climate change. The next week, a consultant emailed the meeting's attendees an eight-page memo titled "Global Climate Science Communications Action Plan," which synthesized what had been discussed in the meeting. "Victory will be achieved," the memo read, when "average citizens 'understand' (recognize) uncertainties in climate science." The memo provided a situation analysis, a series of goals, and three overarching strategies. One, the initiation of a media relations program to generate coverage of scientific uncertainties. Two, the establishment of a "Global Climate Science Information Source" aimed at "raising questions about and undercutting the 'prevailing scientific wisdom.'" And three, the creation of a national outreach and education program aimed at persuading politicians and also teachers and students of their position. Under each strategy, several tactics were listed. To execute them would require immediate funding of $2 million, and millions more in the coming years, the memo stated.

In researching this book, I spoke with two people who attended that 1998 meeting. One of them, Myron Ebell, said he was invited largely because he had worked with a senator who was friendly with API leaders. The other, Steve Milloy, said he was asked to attend because he was a lawyer who had done work on environmental regulation issues. Both men were already established members of the doubt-peddling world. From about 1990 to 1993, Milloy had conducted statistical analyses for a lobbying firm working to fight secondhand-smoke regulation. A few years later, he was tapped to lead The Advancement of Sound Science Coalition, a group founded as a front for Philip Morris that later expanded to advocate for other industries as well (among its advisors were Seitz and Singer). He's most well known for popularizing the term "junk science," which he tends to lob at studies that challenge his political viewpoints. Ebell, too, had connections with the tobacco industry. In the 1990s, with funding from Philip Morris, he promoted "safe cigarettes" and fought proposed tobacco regulation.

Per their recollections, the 1998 meeting took place in API's headquarters in Washington, DC, and Milloy said it lasted perhaps a couple of hours. The industry lobbying group Global Climate Coalition had already been aggressively pushing doubt about climate science for nearly a decade. According to Ebell, the meeting brought together members of that coalition, including representatives from Exxon and Chevron, with people associated with the Cooler Heads Coalition, a recently formed ad hoc group of conservative think tanks working on climate policy. Both Ebell and Milloy recounted the meeting as an informal brainstorming session of sorts. After it ended, the API consultant compiled the proffered ideas into the memo. It "wasn't a group product in the sense that everybody signed onto it and said, 'I agree with everything,'" Ebell said. Rather, he said, it was a conglomeration of ideas and a rough budget that API leadership might fundraise around.

Both men rolled their eyes at the fact that they still get calls about that meeting. The memo was leaked to the *New York*

Times, which "made it a lot more famous and a lot more meaningful than it actually was," Milloy said. Ebell described it as "one small episode" that got undue attention. Plenty of other meetings and other memos transpired around the same time, he said, "but this is the only one that ever became public."

Furthermore, Ebell said, the memo offered nothing beyond what's in the "standard toolbox" for a communications campaign. "When you're waging a political battle, involving persuading the public and elected officials, there's really only so many tools that you have, and there are only so many strategies that are both doable and that might plausibly lead to success," he said. "I don't think there's anything very profound or original or even very creative in it."

Milloy, too, described the meeting as uninspired. "It was just the stupidest meeting I've ever attended," he said. "It was a typical Washington group-grope. People solicit for ideas. People offer ideas up. And nothing really happens for a variety of reasons."

I asked whether the leak had put a stop to the proposed plan. It might have, but other major forces were also at work, they told me. No one felt very fearful that Kyoto would actually be ratified. The year before, the Senate had voted to block any treaty requiring developed nations to limit carbon emissions more than developing nations did—which Kyoto would do. What's more, Milloy said, the industry had begun to split over how to approach the climate question. While some members wanted to keep fighting carbon regulation, others felt regulation was inevitable and trying to stop it was a waste of time, and still others were beginning to see potential profit in alternative energy. This split put API—the group meant to organize action—in an impossible position, Milloy said. "Members start pulling the organization in different directions, and as a result, nothing happens," he said.

Whether nothing actually happened is a matter of debate. The meeting's attendees have long downplayed its significance, as Milloy and Ebell did. But analyses by reporters at the *Guardian* and the watchdog website *DeSmog* concluded that many of the memo's intentions were in fact executed.

One tactic the plan described involved identifying, recruiting, and training a corps of scientists willing to talk to the media about the "climate change debate." Seitz and Singer were already doing this work, but the memo called for "new faces who will add their voices to those recognized scientists who already are vocal." One new face belonged to Willie Soon, a part-time employee of the Harvard-Smithsonian Center for Astrophysics. Soon's education was in aerospace engineering, and in the 1990s he focused his attention on the Sun. He began making the case that it was responsible for global warming. He made media appearances, testified before Congress, and spoke at conferences—labor for which Heartland honored him with a "Courage in Defense of Science Award" in 2014. He was also the speaker who showed the PowerPoint slide about Arctic ice at the 2017 Heartland conference I attended.

An enterprising researcher at Greenpeace realized that because Soon worked for a public institute, his correspondence would be subject to a Freedom of Information Act request. The resulting documents revealed that between 2000 and 2014, Soon received $1.25 million from ExxonMobil, American Petroleum Institute, the coal giant Southern Company, and a foundation founded by the conservative Koch brothers, whose fortune derives partially from oil refining. In at least eight studies, he failed to disclose his funding sources, in violation of ethical guidelines. Soon has insisted that industry funding had not swayed his findings. "No amount of money can influence what I have to say and write, especially on my scientific quest to understand how climate works," he said to a *Boston Globe* reporter in 2013.

Another tactic involved inundating science writers with "a steady stream of climate science information," and newspapers with op-eds and letters to the editor. For a period, the conservative think tank American Enterprise Institute (which received major funding from Exxon) offered scientists and economists $10,000 plus expenses for published articles that contradicted the UN's climate conclusions. Such efforts met with a pliable media market. A 2004 analysis found that 94 percent of stories about global warming published in the *New York Times*, *Washington Post*, *Los Angeles Times*, and *Wall Street Journal* from 1988 to 2002 were out of step with the scientific consensus. A 2020 study of 1,768 press releases issued between 1985 and 2013 found that climate skeptical press releases were twice as likely to receive attention from newspapers compared to those that affirmed the scientific consensus.

Academics who have studied the phenomenon say that the journalistic norm of objectivity left members of the media susceptible to this tactic. Lazy or overworked journalists would give equal time to various "sides" of an issue, even if one side came from a vested interest working to spread disinformation. Audiences likely assumed these debates arose from legitimate scientific disagreement.

Another of the plan's ambitions was to "convince one of the major news national TV journalists (e.g. John Stossel) to produce a report examining the scientific underpinnings of the Kyoto treaty." This goal was accomplished many times over. Stossel, a libertarian-leaning co-anchor of ABC News's newsmagazine *20/20* and later a pundit for Fox Business, produced several segments questioning the link between humans and recent warming. He posted some of them on Stossel in the Classroom, a website that since 1999 has offered lesson plans and learning aids for teachers. As of 2021, one free resource encourages students to explore "both sides" of the "climate debate." Another has students fill in the blanks of a worksheet as they watch a five-minute segment on "environmental fear mongering." The workbook reads "The *New York Times* ______ reporting reads like a Greenpeace newsletter." (Answer: climate.) "No dissent allowed when you're teaching your children about global warming. They're going to be told that there's one view, and the view is that we face a climate danger and that we ______ act." (Answer: must.) Stossel in the Classroom boasts that its resources have been used by 150,000 teachers.

The 1998 memo explicitly stated its desire to reach classrooms. "Informing teachers/students about uncertainties in climate science will begin to erect a barrier against further efforts to impose Kyoto-like measures in the future." They were to form a "Science Education Task Group," which would "distribute educational materials directly to schools and through grassroots organizations." The program was to launch right away. We don't know if the Science Education Task Group was ever established, or if it, too, was derailed by the media attention or other factors.

But we do know that in the following decade, several of the meeting's participants and their allies devoted time and resources to classroom campaigns, such as those described in chapter 6.

According to the action plan, the Science Education Task Group would serve as the point of outreach to National Science Teachers Association to collaborate on school materials. The membership of the NSTA (which has since changed its name to the National Science Teaching Association) stood around 55,000, making it the largest group of science educators in the world. Like many professional organizations, it has long been supported by corporate sponsorship. In 1998, as today, some of those sponsors were fossil fuel organizations. The leadership of the NSTA has told me that at no point did they allow their funders to influence their content. But at least three resources developed in the years after that meeting either included links to denial organizations or exaggerated doubts about climate change.

The first instance appeared in 1998, when the NSTA launched a website for teachers and students about the science of energy, funded by the American Petroleum Institute. The page is now defunct, but while it operated, it provided cursory information about renewable energy and encyclopedic information about petroleum—its history, extraction, refinement, transportation, and uses. It only mentioned climate change in passing—describing it as "the buildup of carbon dioxide some fear is warming the Earth's atmosphere"—but among the resources it directs educators to is the Global Climate Coalition, the doubt-pushing industry group.

In 2003, the NSTA collaborated with ConocoPhillips on a ten-part series of short science films for middle- and high-schoolers. The series, called *The Search for Solutions,* lists NSTA's executive director and ConocoPhillips's director of corporate affairs as executive producers. The film series discusses climate change extensively in five of its ten episodes. But most notable is what isn't discussed: the mechanism by which the climate is changing, the extent of that change, or fossil fuels' contribution to it. Instead, one segment is devoted to natural climate changes, another to the fallibility of climate models, and another to the possibility that recent climate change could be caused by both humans and nature. The scientists who appear in the film are quoted in a way that inflates uncertainty. They repeatedly say things like, "The climate has always changed, it always will change, there have been dramatic changes in the past, and there will be dramatic changes in the future." Which is true, but glosses over the extraordinary situation we find ourselves in at present. Another scientist, referring to the conclusions of climate models, says, "We have to look at the answer and say, okay, how much do we believe it?" but the film moves on before he answers that question. An accompanying teacher's guide asserted, "Some scientists believe that the high level of present day CO2 will soon be, if not already, on the decrease." Two years earlier the National Academies of Science had issued a blunt report affirming that whatever uncertainty remained in the models, their overall conclusion had already been proven right. "Greenhouse gases are accumulating in earth's atmosphere as a result of human activities," it read. "Temperatures are, in fact, rising."

A third instance appeared in a book called *Resources for Environmental Literacy: Five Teaching Modules for Middle and High School Teachers,* published by the NSTA in 2007 and still for sale on their site as of 2020. The book was a collaboration between the NSTA and the Environmental Literacy Council, an organization established around 1997 by a man named Jeffrey Salmon—who, as it happened, attended the 1998 meeting. Salmon already possessed a long list of conservative credentials and has amassed more since: He worked as a speechwriter for then-senator Dick Cheney, ran the conservative think tank George C. Marshall Foundation, held a job in the Bush Department of Energy, and today helps lead the CO2 Coalition, whose raison d'être is to persuade the world that "additional CO2 will be a net benefit" for society. The book includes a module on global climate change, which begins, "Most scientists believe that the Earth's climate is changing and is in fact heating up. However, there are considerable differences among the views with regard to the rate of change, the impact on our environment, and what can and should be done about it." The module suggests that "some scientists" believe that the cause of recent warming is "a repeating 1,500-year solar irradiance cycle." Teachers should take no stand on the extent of global warming, the book insists. "It is very important that students learn to draw their own conclusions."

Controversy over NSTA's relationship with fossil fuel interests erupted in 2006, when Laurie David, a producer of Al Gore's *An Inconvenient Truth,* wrote in a *Washington Post* editorial that her company had offered to give the NSTA 50,000 free copies of the film to distribute to teachers. She wrote that the NSTA had rejected the offer because it feared doing so would be seen as a "political endorsement." Furthermore, the NSTA had told her that distributing the film could place "unnecessary risk" on its fundraising efforts.

The editorial scandalized the organization's membership, who wrote hundreds of emails to its leadership on the issue. In response, Executive Director Gerry Wheeler released a flurry of press releases flatly contesting the characterization that the NSTA "eagerly pushes corporate messages about the environment." The NSTA did accept corporate support—to the tune of 16 percent of its budget—but never sold editorial control, he asserted. To further quell members' concerns, the NSTA board convened a blue ribbon panel of scientists and educators to investigate. Several months later, the panel issued a report saying it had found no evidence of corporate influence on content, but that the NSTA needed clearer policies. Current NSTA communications staff said the ConocoPhillips films counted among the materials the panel reviewed, but they were unsure if the NSTA's 1998 website on the science of energy did.

When I spoke with Wheeler, who led the organization for thirteen years before retiring in 2008, he said that he himself had seen Al Gore speak multiple times and "completely agreed with him" about the climate crisis. Worried that climate science was not adequately taught in American classrooms, Wheeler had invited climate scientists to speak in prominent slots at the organization's popular conferences. He reiterated that at no point under his watch did the organization allow fossil fuel sponsors to sway the materials the NSTA produced.

I told Wheeler about the API's 1998 meeting and the action plan that targeted the NSTA to develop climate denial materials. He said it was news to him, but he wasn't entirely surprised. He had been invited to some meetings with API in that era and "quickly got out" because it became clear what their "game" was: "They were definitely keen on downplaying climate change." It made sense to him that they would try to target the NSTA, which had "the big pipe" to science teachers. "I'm not naive. I know that ExxonMobil and all those others [sponsor the NSTA] partly because they get a tax break, and partly because they want to be seen as good citizens. It's propaganda from their point of view. I certainly appreciate that. But what I worried about mostly—entirely, actually—was that we didn't become a vehicle for their messaging." When I brought up the API-funded site on energy, Wheeler said he recalled nothing about it and lacked an explanation. "I'm embarrassed that that happened," he said. "Some of that stuff comes into being just because it's a big enough organization that some things slip through the cracks." I asked current NSTA officials if the organization kept archives that might help elucidate the editorial process for the three suspect projects, but they said they were unable to find any.

When Heartland sent out materials to science teachers in 2017, the NSTA responded forcefully, using multiple mediums to inform its members of the disinformation campaign. The next year, NSTA leadership finally issued a statement standing by the legitimacy and conclusions of modern climate science and calling on educators to teach those conclusions. When I asked about the older materials, the NSTA's recently selected executive director, Erika Shugart, sent the statement, "NSTA supports quality science education and we fully embrace the teaching of climate science and climate change. We must have accurate science in the classroom and if the science in our resources is not accurate, then we correct it." As of 2021, *Resources for Environmental Literacy* appears to have been removed from the organization's online bookstore.

The industry members that spent the 1990s and early 2000s pushing doubt about climate science backed off those messages by the 2010s. ExxonMobil was one of the last holdouts, but in 2007, decades after its own scientists concluded that climate change could affect its profitability, it disclosed that conclusion to shareholders. Even the American Petroleum Institute in recent years has voiced support for the goal of reducing greenhouse gases in the atmosphere.

But in their place remains an ecosystem of climate denial groups willing to push the doubt that mainstream companies no longer will. The Heartland Institute is at the center of this community. Others include the Competitive Enterprise Institute (where Ebell works), the CO2 Coalition, the Committee for a Constructive Tomorrow, the Cooler Heads Coalition, the American Enterprise Institute, Center for the Study of CO2 & Global Change, Science and Environmental Policy Project, Milloy's JunkScience.com, Germany's EIKE, and Canada's Fraser Institute. These groups distinguish themselves by the temerity with which they are willing to attack legitimate science.

Heartland may not have attended the 1998 meeting, but it had—or soon would have—relationships with many of the people and organizations that did. Eventually it would be doing yeoman's work in support of the action plan's goals. Since its establishment in 1984, Heartland has supported an assortment of free-market causes—lowering taxes, reversing regulatory oversight, and promoting school choice. It has received funding from the tobacco industry—and to this day asserts that the health risks of smoking have been exaggerated. One of Heartland's main products is a series of public policy newspapers it sends to conservative state legislators; the group has claimed that 78 percent of state elected officials read one or more of their newspapers.

Climate change has been a primary thrust of the organization's work since the late 1990s, and they received at least $650,000 from ExxonMobil before the company walked back their support of climate denial groups. In 2008, Heartland began hosting its regular "International Conference on Climate Change." It has published dozens of reports and books with titles like, *Is the U.S. Surface Temperature Record Reliable?* and *The Neglected Sun*, as well as hundreds of articles amplifying the voices of people who reject the consensus on climate change. The 1998 action plan prescribed developing an alternative to the UN's reports, so as to provide a "complete scientific critique of the IPCC research and its conclusions." In 2009, Heartland published the first in a series of volumes called *Climate Change Reconsidered: The Report of the Nongovernmental International Panel on Climate Change (NIPCC)*, coauthored by Singer and marketed as an "880-page rebuttal" to the UN IPCC's analysis. The action plan called for "advertising the scientific uncertainties in select markets." In 2012, Heartland experimented with a billboard campaign pushing climate doubt. This ended abruptly after backlash over a billboard depicting the "Unabomber," Ted Kaczynski, along with the words, "I still believe in global warming. Do you?"

In 2012, scientist and activist Peter Gleick, using false pretenses, obtained and shared with media confidential internal Heartland documents that, in addition to providing insights into the organization's funding (a single anonymous donor had contributed $13.3 million over five years, two-thirds of which went to climate programs), described their interest in messaging to children. "Many people lament the absence of educational material suitable for K-12 students on global warming that isn't alarmist or overtly political. Heartland has tried to make material available to teachers, but has had only limited success," states a strategy document. (Heartland claims this document is fake, but an external review found no evidence of its falsification.) Fortunately, it said, their anonymous donor had pledged $100,000 for the creation of modules on climate topics, such as the "major scientific controversy" of whether humans are to blame for global warming. These modules never appeared. But Heartland's focus on getting its messages to classrooms did not wane. Heartland had mailed *Unstoppable Solar Cycles* and other curricular materials to 11,250 schools in Canada in 2008. The next year, it sent the denialist pamphlet *The Skeptic's Handbook* to the presidents of 14,000 public school boards. It sent a small run of *Why Scientists Disagree About Global Warming* to science teachers in 2016, and and the next year vastly expanded that campaign with the one Bast told me about. They have also sent full volumes of their *Climate Change Reconsidered* series, roughly 4,000 pages of material, to public and school libraries.

More than twenty years after the 1998 meeting, Milloy and Ebell are still both deeply involved in the climate denial world. Milloy is a Fox News contributor. Ebell boasts in his bio at the conservative think tank Competitive Enterprise Institute that he was featured in Greenpeace's "Field Guide to Climate Criminals," among other "recognitions" of his work. And, with the election of President Donald Trump in 2016, both achieved some measure of power: They worked on his transition team for the US

Environmental Protection Agency in 2017, and Ebell led efforts to persuade Trump to pull out of the UN's Paris climate accord.

Both Ebell and Milloy see their movement as an underdog fighting against a machine operated by corrupt environmentalists and self-serving scientists. Ebell said that there "used to be a hundred-to-one disparity in resources and money" between the movement pushing for action on the climate crisis and the one pushing against it. "I'd say it's more like a thousand-to-one now."

But if there is a shortage of funding in the climate denial business, it's a recent one. Drexel University sociologist Robert Brulle tallied the budgets of ninety-one think tanks, advocacy organizations, and trade associations that comprised the climate denial industry between 2003 and 2010 and found they had collectively been funded $900 million a year. Brulle's team also sorted through 64,162 lobbying reports from 2000 to 2016 and found that $2 billion was devoted to persuading the US Congress against action on the climate crisis—ten times as much as environmentalists spent persuading them toward it. The spending peaked in 2009 as the Democratic-controlled House of Representatives considered cap-and-trade legislation. The bill died.

While the 1998 action plan wasn't necessarily executed as planned (at least publicly), its mission found considerable success. Not only were barriers erected "against further efforts to impose Kyoto-like measures in the future," but nearly twenty years later, the political climate was such that Trump could hire a slew of climate deniers, scrub mention of the climate crisis from the EPA's website, and reverse US commitments to cutting emissions—which had been modest to begin with. Climate denialism became and remains a pillar of conservative beliefs. The Republican stance has overwhelmingly been either to deny human-caused climate change or eschew substantive action on it; Democrats are more likely to accept the science, but have also hesitated to adopt serious changes.

The American public's perspective on global climate change remains wildly out of step with that of scientists. As of 2019, 30 percent of Americans falsely thought global warming was mostly natural. Four in five didn't know there was scientific consensus on the question. This should come as no surprise, said Niepold. "We are the only country in the world that has had a multi-decade, multibillion-dollar deny-delay-confuse campaign," he said. The idea that the science is debatable didn't occur naturally, but rather was "implanted and sustained."

Tendrils of connection, if not causation, link monied efforts and what children learn about climate change today. A third of science teachers tell students that "many scientists believe" global climate change is natural. Many mainstream textbooks continue to write about global warming in stilted language that communicates uncertainty. Some states hold onto science standards that ignore or misrepresent climate science. Every year, state lawmakers propose legislation that would allow teachers to miseducate their students about it. Troves of misleading educational materials are available online.

Had those who wrote the 1998 memo been asked what they hoped American climate education would look like more than twenty years later, they might describe this exact reality. Victory, one might say, has been achieved.

Epilogue

Something had been pestering Nakowa Kelley, and as he and his friends settled into their seventh-grade science class, he finally said it aloud. “This global warming stuff? My parents said it’s not true.”

His science teacher, Marc Kessler, had been expecting this. “So you’re getting mixed messages,” he said. “That must be a little challenging.”

The twelve-year-old recited the arguments he’d heard at home: If the earth is warming, why had it snowed so much that winter? And without carbon dioxide we’d be dead, so what was wrong with having a little more? On the other hand, the class had spent days looking at NASA data, which seemed convincing, too.

“I don’t know who to believe,” Nakowa said miserably, and then confessed, eyeing Mr. Kessler, “My parents also told me not to argue with the teacher.”

Mr. Kessler had navigated this discussion before. He had been teaching middle-school science for years in Paradise,

California, a low-income, predominantly white, and politically deep-red community about fourteen miles into the Sierra foothills above my hometown of Chico. Every year, he taught a unit on climate change. Every year, students told him they'd heard it was a hoax.

"It's super respectful of your parents to say not to argue with the teacher, but it's totally okay in my class to bring up differing views," he said. "If scientists didn't argue, we wouldn't get to the truth."

Getting to the truth about climate change has proved difficult for many American children, but all over the country there are teachers like Mr. Kessler helping them get there. Mr. Kessler devotes time each year to climate science not because California state standards require him to, but because he believes his students deserve it. This is also true of Ms. Del Real at my alma mater, Ms. Lao in Oklahoma, and many others I met while researching this book. These teachers exist not just in cities and college towns, but in the nation's reddest states and also in the reddest parts of blue states. Their work helps children understand the climate crisis unfolding around them and prepares them to participate in civic deliberation over what to do next.

Not all teachers are as adept or motivated as Mr. Kessler. Some don't teach climate science because they reject its conclusions or fear controversy. But research shows that even teachers who accept the science often do a subpar job of teaching it, particularly if they don't have a good grasp of it themselves.

A raft of environmental and education organizations have set out to address this problem through professional development programs and other resources. Washington State has allocated millions of dollars since 2018 to providing climate education training to teachers; one in five teachers in the state took part in just its first two years. The National Center for Science Education has developed lesson plans aimed at navigating climate misinformation. The Climate Literacy and Energy Awareness Network, the group that surveyed 30,000 free climate education resources and found only 700 acceptable, has amassed those in the latter category into a one-stop portal. The publishers of a book called *The Teacher-Friendly Guide to Teaching Climate Change* mailed it to 50,000 teachers in forty-three states—a campaign meant to serve as an antidote to Heartland's materials.

One innovative program led by the nonprofit Alliance for Climate Education sidesteps a reliance on teachers by hosting high-school assemblies that use performance, animation, and storytelling to connect directly with students on the subject. This approach is clever, said NOAA's Frank Niepold, because if students get excited about climate science, they put pressure on the education system from the inside. "Teachers are always looking for ways to engage students, so they're like, 'Oh, you're curious about that? That's important to you? Let's do it,'" Niepold said.

When given the chance, young people *are* curious about it, and *do* consider it important. A 2021 Pew survey investigated each generation's attitudes about global warming. Gen-Z and Millennial respondents voiced considerably more concern about climate change and more willingness to do something about it than older respondents. This generation gap was especially large among Republicans—Gen-Z Republicans were three times as likely as Baby Boomers and older Republicans to support phasing out fossil fuel use entirely. Of all the generations,

Gen-Z respondents were most likely to say that addressing climate change was their top personal concern. They also reported seeing more content on social media about action on climate change than their elders.

School is not the only—or even the primary—place that young people learn about the climate crisis. Youth are likely to encounter the subject in museums, zoos, and parks; on television, films, and social media; at church; on their phones; and around the kitchen table. But schools occupy a special place in our communities as the only institutions that reach virtually all young people. If the school system is influenced by adult politics to provide an uneven education about the climate crisis—so students living in some places receive a high-quality education while those living elsewhere receive misinformation—inequities arise.

What does a quality classroom education about the climate crisis look like? Science educators agree that, at minimum, students should enter the adult world knowing the so-called "big five facts" about the crisis: It's real. It's us. Experts agree. It's bad. There's hope.

The details of how these concepts are taught matter. A 2017 review of forty-nine studies on climate curricula found the most effective made explicit connections between the global phenomenon and its locally relevant and meaningful effects. They also employed techniques that engaged learners, favoring labs, field trips, role-plays, small-group discussions, and inquiry-based learning over lectures or memorization. Some climate educators have incorporated indigenous knowledge systems into their lessons. Ecological knowledge was passed generation to generation on this continent long before colonizers established their first school here, and those traditions have persisted and evolved. Indigenous communities have diverse approaches to climate education, of course, but broadly speaking, they tend to be place-based, observational, and participatory methods of studying the interconnections between humans and our environs.

When Niepold asks young people what they want more of in their climate education, they overwhelmingly have one request: solutions. "Right now, climate education is maybe 99 percent problem, 1 percent solution. Young people want 20 percent problem, 80 percent solution," he said. Rather than scaring children about their future, a good climate education should leave children "fired up and ready to roll up their sleeves and get to work, because we've got some stuff to do."

There is tension among science educators about whether a goal of climate education should be to inspire action. Some argue that such a goal turns classrooms into a political space. But that argument assumes schools are a politically neutral ground to begin with. It also pretends that the climate crisis is simply another body of knowledge for students to understand, rather than something relevant to a learner's personal life and future. "We're not talking about how kids understand mitosis, right?" said researcher K. C. Busch. "We're talking about a topic that has a real-world, immediate implication."

Busch herself was a teacher for twelve years before joining academia. "As a teacher, my goal was never to make students who could reproduce knowledge for a test," she said. "The goal was to produce people who could go out and live their lives and contribute to their communities. It was to produce citizens. And citizens have to take action."

On the whole, science remains one of the most trusted institutions in America, and that trust has remained relatively stable for the last forty years. Dig into those numbers, however, and patterns appear. Christians have less trust in science than nonreligious people, rural dwellers less than urbanites, and Republicans less than Democrats. There have been times in the last forty years when those camps had equal confidence in science. Today, the space between them is as large as it's ever been.

These patterns are no accident of history. Rather, they are the product of successful disinformation campaigns, animated not by science but by ideology. Those who lead these efforts are united in a singular belief in the free market coupled with a profound distaste for government regulation. They hold up the free market as an unimpeachable source of American prosperity. But as scientists have studied the planet's ecosystems, they have again and again arrived at the conclusion that unregulated markets have inflicted environmental costs that they don't pay for, causing real and lasting harm to humans, animals, and plants. Acknowledging that truth would concede that unfettered capitalism is a deeply flawed system. Instead, free-marketeers attack the messenger—science. By undermining the science on tobacco, acid rain, asbestos, nuclear weapons, ozone depletion, and climate change, they helped birth the "post-truth" world we find ourselves in now, where even the most authoritative sources are met with suspicion and can be contradicted with "facts" of dubious origin.

The consequences of this mistrust played out fatally during the COVID-19 pandemic. After the initial spike, case numbers grew most precipitously in parts of the country that scorned expert advice on masks, and then on vaccination. The anti-science movement found a powerful ally in President Trump, who considered the warnings of experts like his own chief immunologist, Dr. Anthony Fauci, as impediments to his beliefs about the world. Rather than adjust his beliefs, he adjusted his estimation of Dr. Fauci. The right wing doesn't lay sole claim to this kind of thinking. Fears about childhood vaccines and the health impacts of genetically modified crops arise across the political spectrum despite a lack of scientific support, for example. Scientists and their allies have not yet proven capable of combating this hostility, of laying such fears to rest in a meaningful way.

In this context, children can be forgiven if, rather than sort out what is true and what isn't, they defer to what the most important people in their lives think of climate change, even as they are already being licked by the first flames of the climate crisis.

On a brisk November morning a few months before I met him in Mr. Kessler's class, twelve-year-old Nakowa stood in his driveway, sniffing. He walked inside and told his mom it smelled like fire. "Oh, honey, don't worry about that," Nancy Kelley told him. "Somebody's probably burning their leaves." Nakowa looked again. Debris floated down like a rain of dusky feathers. "No, Mom. There's great big cinders in the driveway!" They turned on the news and learned there was indeed a fire in Paradise, but their neighborhood—labeled zone 10 on the town's fire map—was not on the evacuation list, so they went about their morning as usual. "The TV kept saying, 'Zone 10 is safe.

Zone 10 is safe,'" Mrs. Kelley told me. But fifteen minutes later, a police officer stood in front of their house, shouting, "Get out! Get out now!"

In one respect, the Kelleys were lucky. They lived on the lower end of town, so they were toward the front edge of a traffic jam 52,000 people strong. Nonetheless, it took them two hours to cover the fourteen miles from Paradise to Chico. They watched the blue sky turn night black in their rearview mirror, and then all around them. Flying embers converted the dry grass alongside their van to flames. They escaped, but miles behind them, people burned alive in their cars. The Kelleys' home and thousands of others went up in smoke.

When I met the Kelleys, the devastating Camp Fire had long been extinguished, but every aspect of their life remained touched by it. The Kelleys and their three school-aged boys were living in a one-bedroom apartment. Beds congested the bedroom. Dressers lined the living room. Nakowa's dad had converted the coffee table into a woodworking station. A pair of parrots squawked from their perch in the kitchen. Nakowa's school, Paradise Intermediate, had relocated to the only real estate available in suddenly overcrowded Chico: a shuttered hardware store in a strip mall. Nakowa and his 200 classmates were served lunch at a checkout counter and played freeze tag in the garden center. Mr. Kessler's class occupied aisles 9 and 10, where you used to find ceiling fans and light fixtures.

When Nakowa raised his confusion over who to believe about climate change, Mr. Kessler hushed the class. "So, could I have everybody not talk for just a second?" he asked. His students complied, but it did little to quell the clamor. The aisle shelves framing the classroom were six feet tall, while the metal roof rose more than twenty. Even as Nakowa and his classmates sat mutely, they could hear every word uttered in aisles 6 and 7—math class—and 12 and 13—history class. To help his students hear him, Kessler wore a microphone around his neck. Now he lifted it and spoke carefully. "Sometimes you will learn things that conflict with what you're hearing somewhere else," he said. "My job is to provide you with the best scientific data that's out there, and then we interpret and predict. This isn't me telling you what you should believe; it's just discovery. That's what science is."

Mr. Kessler's response did not put Nakowa's mind at ease. Three days later, as the climate change unit was wrapping up, Mr. Kessler gave his students a writing prompt: How has climate change affected your life so far? And what effects do you think climate change will have on your life in the next fifty years? "It hasn't effected me life at all yet," wrote the boy whose home had burned down five months earlier. "I don't know if it will do anything to my life in fifty years because I don't know if i believe it yet."

The first person to see this project as worthy of a book was Raney Aronson-Rath, executive producer of *FRONTLINE*. In addition to gifting me with her confidence and enthusiasm, she saw to it that *FRONTLINE* and WGBH provided me with crucial material support for this reporting.

I am indebted to pretty much everyone who worked at *FRONTLINE* from 2015 to 2021, and I'll highlight just a few who directly touched this project. Michelle Mizner helped conceive of and give shape to this reporting. She was the book's first and best reader. I hope we get to tell many more stories together. Ben C. Solomon accompanied me and Michelle to Oklahoma in February 2020 in hopes of making a film about climate education. The pandemic aborted that film project, but I'm grateful for his work and excellent company. In May 2019, Sophie McKibben traveled with me to Chico to gather tape for a podcast on the subject. I learned so much about audio journalism in just those few days and would happily accompany her on a reporting trip anywhere, anytime. The unflappable and kind Jason Breslow did a lot of the editing on my early reporting on this story. At one point, Abby Johnston applied her penetrating editorial eye to an unreadable sixty-page reporting memo I had written and provided astute feedback. Patrice Taddonio helped think through promotional copy and live-texted her reactions to the manuscript, to my great delight. Whenever I needed to talk through a sticky reporting problem—which was often—Sarah Childress made her prodigious brain available to me.

The GroundTruth Project's support dates to this book's inception. Not only did they help pay my salary for a period, they funded the trip that Michelle and I took to the Marshall Islands in 2017, which resulted in the Emmy-winning interactive film *The Last Generation*, and led to the reporting questions motivating this book. GroundTruth also sent me to two United Nations Climate Change Conferences, where I developed important relationships and a much broader understanding of the crisis, its politics, and its solutions. The organization's visionary founder, Charlie Sennott, has been a spirited cheerleader of this project. Marissa Miley took about a thousand of my calls over the last four years and has been a critical editorial resource and friend to me. GroundTruth also allowed me to bogart the labor of intern Jake Friedman, who never complained about the tedious assignments I gave him.

The project was also meaningfully supported by Marquette University's O'Brien Fellowship in Public Service Journalism, where I spent nine months reporting. Thanks to the fellows of the 2018–2019 class and to the program's director, all excellent journalists who were generous with their editorial advice and encouragement. The work of student journalists Robin Di Giacinto, Colleen DuVall, John Hand, Clara Janzen, and Lucie Sullivan is all over this book. John signed up for a summer internship at *FRONTLINE,* where he reported the hell out of Idaho's extraordinary science standards drama. Lucie Sullivan did as well, and gifted us all with her stellar research and wit.

The good folks at Columbia Global Reports treated this book with professionalism, generosity, and patience. And this book required so much patience. I am grateful to Jimmy So for his deft and efficient edits, to Camille McDuffie for her persistence, knowledge, and for forgiving my trespasses against a number of deadlines, and to Nick Lemann for his longtime mentorship. Nobody would know about this book without Megan Posco's megaphone; I am fortunate to have her on this team.

Special thanks to all the educators and learners who generously gave me their time, expertise and stories, including Kristen Del Real, Barbara Forrest, Melissa Lau, Jonathan Kahl, Nakowa Kelley (and family), Marc Kessler, Izerman Yamaguchi-Kotton (and family), Paige Miller, Tiffany Neill, Charles L. Nokes, Lori Palmer, Leslie Pitman, Andrea C. Sampley, and Waisake Savu. I also extend gratitude to the many educators and learners whose stories do not appear in this book, because their insights certainly do. The encyclopedic historical knowledge of Glenn Branch of the National Center for Science Education made parts of this book much more robust than they could have been otherwise. Eugenie Scott went out of her way to dig up decades-old documents for me.

I am lucky to have a community that constantly fills my world with joy, hilarity, sanity, comfort, gut checks and assistance. Jenni Monet, from whom I've learned so much about journalistic integrity, listened to a portion of this book during a cross-country road trip in the middle of the pandemic. Dr. Misty Parker also read many chapters and gave invaluable feedback and support throughout. I am glad to be sharing this life with her and her big brains and heart. I have so much gratitude and love in my heart for Phoebe Ambrosia, Bruce and Tamara Barak Aparton, Aaron and Isabel Casillas, Aneel Chima and family, Jolina Clement and family, Celeste Doaks and Karl Henzy, Maria and Jael Garcia, Samatha Lewis and David Cohen, Victor Morales, Paulo Olivares and Erica Kiely, Laurel Paget-Seekins and Alexandra Bertran, Alicia and Evan Rusoja, Jesse Silacci and Hanni Liliedahl, Craig and Rico Lindeman Smith, Loreto Soto Bustos, Will and Tammy Jo Anderson Taft, John Upton, Shelley Wright and Quinn Konopacky, and Max Zachai and Jethro Smith. My life in Boston wouldn't be any good without Lian Hua Barry and her wonderful clan, Anna Pierce, Alecia Robinson, Emma Varsanyi (who took the author photo), my FRONTLINE pals, and, of course, my beloved Adventurbu Crew fam.

The love and generosity of spirit abundant in both my Worth and Pryor families is a wonder to behold, and I have had the privilege of witnessing that wonder all my life. I owe a lot to a few people who are gone. My late aunt Kristyna Demaree, who shined so much light on my life. Judy Zachai, who saw and believed in me. My grandfather Chuck Worth, who gifted me with his memoirs. And my cherished grandmother Jeanie Reyes, whose lifelong evolution and radically honorable values still guide me. Once, she asked me if I thought I'd ever write a book. I said nah, I doubt I could write a good book. "So write a shitty book!" she said. This is the single best piece of writing advice I have ever received and I have relied on it heavily.

Finally, this book is dedicated to my parents, Denise and Chuck Worth, who among many other things, taught me the value of a meaningful career, and what it looks like to show up for the people you love. And to my late brother Kit, who made me who I am.

FURTHER READING

For educators in search of resources on how to teach children about the climate crisis, the Climate Literacy and Energy Awareness Network has surveyed 30,000 of them and has amassed the most accurate and up-to-date in a one-stop portal. Find it at www.cleanet.org.

To understand how a handful of politically motivated scientists helped sway Americans' understanding of global warming, read Naomi Oreskes and Erik M. Conway's brilliant book *Merchants of Doubt*. The book's last chapters have an especially incisive analysis of the relationship between such doubt campaigns and unmitigated capitalism. To learn about the methods by which moneyed interests dictate public policy and opinion, pick up Jane Mayer's *Dark Money*, which describes them in crushing detail.

For essential (if infuriating) reading on what Exxon knew about the crisis and when they knew it, check out the series of stories that earned *Inside Climate News* a finalist nomination for the 2016 Pulitzer Prize in public service journalism, along with the reporting in the same vein published that year by the *Los Angeles Times* and the Energy and Environmental Reporting Project at Columbia Journalism School.

Though dated, Sheila Harty's book *Hucksters in the Classroom: A Review of Industry Propaganda in Schools* gives a thorough history of private interest efforts to market to children. For more on how the oil industry has done so in recent years, read Jie Jenny Zou's stellar reporting in "Oil's Pipeline to America's Schools" for The Center for Public Integrity. Asher Price's investigation for the *Austin American-Statesman*, "How a Natural Gas Group Pushed for New Energy Curriculum in Texas," provides an example of how industry messaging to schoolchildren is sometimes paid for by taxpayers.

Robin Globus Veldman's excellent book, *The Gospel of Climate Skepticism: Why Evangelical Christians Oppose Action on Climate Change*, and Lydia Bean and Steve Teles's report, *Spreading the Gospel of Climate Change: An Evangelical Battleground*, together tell the fascinating story of how climate denial became a pillar of Evangelical beliefs even as forces within the church pushed for the crisis to be taken seriously.

K.C. Busch of North Carolina State University has tackled the climate education question from more angles than anyone else in the field. Her research is extensive, illuminating, and worth reading.

The National Center for Science Education closely follows developments in climate and evolution education across the country. Their site (ncse.ngo) and newsletter are the best ways to track the bills proposed each year that try to alter what kids learn about science.

NOTES

INTRODUCTION

12 **school day in America:** National Center for Education Statistics, Current Digest Tables, nces.ed.gov/programs/digest/current_tables.asp. Accessed August 2021.

13 **linking cigarettes and cancer:** Douglas Fischer, "Climate Risks as Conclusive as Link Between Smoking and Lung Cancer," *Scientific American*, March 19, 2014.

13 **a 2021 UN survey:** "The People's Climate Vote," United Nations Development Programme (UNDP) and the University of Oxford, January 2021.

14 **unleashed natural disasters:** Adam B. Smith, "2010-2019: A Landmark Decade of U.S. Billion-Dollar Weather and Climate Disasters," Climate.gov, January 8, 2020.

14 **more megafires:** Ed Struzik, "The Age of Megafires: The World Hits a Climate Tipping Point," YaleEnvironment360, September 17, 2020.

14 **most destructive fire:** "Top 20 Most Destructive California Wildfires," CalFire, www.fire.ca.gov/media/t1rdhizr/top20_destruction.pdf.

14 **covered with its fingerprints:** Matthew Cappucci and Jason Samenow, "The Weather and Climate Behind the California Infernos That Wrecked Paradise and Torched Malibu," *Washington Post*, November 12, 2018.

14 **wildfires will strike:** "California's Fourth Climate Change Assessment," January 16, 2019, www.climateassessment.ca.gov/state/.

14 **below the high tide line:** "Flooded Future: Global Vulnerability to Sea Level Rise Worse Than Previously Understood," Climate Central, October 29, 2019.

15 **followed students:** Eugene C. Cordero, Diana Centeno, and Anne Marie Todd, "The Role of Climate Change Education on Individual Lifetime Carbon Emissions," Public Library of Science, February 4, 2020.

15 **2008 treatise**: The essay, "What are Children Being Taught in School about Anthropogenic Climate Change?" by Kim Kastens and Margaret Turrin was published on p. 48 of the report by Bud Ward, "Communicating on Climate Change: An Essential Resource for Journalists, Scientists, and Educators," Metcalf Institute for Marine & Environmental Reporting, 2008.

15 **study of middle-school children**: Danielle F. Lawson et al, "Children Can Foster Climate

Change Concern Among Their Parents," *Nature*, May 6, 2019.

16 **skipped school:** Damian Carrington, "School Climate Strikes: 1.4 Million People Took Part, Say Campaigners," *Guardian*, March 19, 2019.

16 **sued the federal government:** John Schwartz, "Young People Are Suing the Trump Administration over Climate Change. She's Their Lawyer," *New York Times*, October 23, 2018.

16 **trillions of dollars:** "Unburnable Carbon—Are the World's Financial Markets Carrying a Carbon Bubble?" Carbon Tracker, 2011.

CHAPTER ONE

19 **no more than three eggs per week:** Donald J. McNamara, "The Fifty Year Rehabilitation of the Egg," *Nutrients*, October 21, 2015.

20 **far more complicated:** Jessica Brown, "The Truth About Eating Eggs," BBC, April 23, 2020.

20 **study examining 12,000 peer-reviewed papers:** John Cook et al, "Quantifying the Consensus on Anthropogenic Global Warming in the Scientific Literature," *Environmental Research Letters*, May 15, 2013.

21 **the 3 percent that disagreed:** Rasmus E. Benestad et al, "Learning from Mistakes in Climate Research," *Theoretical and Applied Climatology*, August 20, 2015.

21 **A review of global warming papers:** James Powell, "Scientists Reach 100% Consensus on Anthropogenic Global Warming," *Bulletin of Science, Technology & Society*, November 20, 2019.

21 **nearly 50 percent:** Rebecca Lindsey, "Climate Change: Atmospheric Carbon Dioxide," climate.gov, August 14, 2020.

21 **first floated in the 1820s:** Peter Lynch, "How Joseph Fourier Discovered the Greenhouse Effect," *Irish Times*, March 21, 2019.

21 **changing the proportion of those ingredients:** Steve Graham, "John Tyndall (1820–1893)," NASA Earth Observatory, October 8, 1999.

21 **The degree to which that was true:** Ian Sample, "The Father of Climate Change," *Guardian*, June 30, 2005.

22 **"indulge in the pleasant belief":** Henning Rodhe, Robert Charlson, and Elisabeth Crawford, "Svante Arrhenius and the Greenhouse Effect," *Ambio*, February 1997.

22 **"likely to prove beneficial":** Leo Hickman, "How the Burning of Fossil Fuels Was Linked to a Warming World in 1938," *Guardian*, April 22, 2013.

22 **"may be sufficient":** "Restoring the Quality of our Environment: Report of the Environmental Pollution Panel," President's Science Advisory Committee, 1965, accessed via Climate Files.

22 **"this generation has altered the composition of the atmosphere":** "Special Message to the Congress on Conservation and Restoration of Natural Beauty, February 8, 1965," Public Papers of the Presidents of the United States: Lyndon B. Johnson, 1965.

22 **a team drilled a cylinder:** J. M. Barnola et al, "Vostok Ice Core Provides 160,000-Year Record of Atmospheric CO2," *Nature*, October 1, 1987.

23 **more than 350 ppm for the first time:** Mauna Loa monthly carbon dioxide measurements, www.co2.earth/monthly-co2.

23 **"Come back in forty-nine years":** Naomi Oreskes and Erik M. Conway, *Merchants of Doubt* (Bloomsbury Press, 2010), p. 173.

23 **certain of the following:** "Policymaker Summary of Working Group I (Scientific Assessment of Climate Change)," IPCC First Assessment Report, 1990.

23 **five hottest years in human history:** "2020 Tied for Warmest Year on Record, NASA Analysis Shows," NASA release 21-005, January 14, 2021.

24 **oceans' five hottest years:** Lijing Cheng et al, "Record-Setting Ocean Warmth Continued in 2019," *Advances in Atmospheric Sciences*, February 2020.

24 **ice sheets covering Antarctica and Greenland:** NASA, climate.nasa.gov/vital-signs/ice-sheets/.

24 **Sea levels have risen:** NASA, climate.nasa.gov/vital-signs/sea-level/.

24 **California's summertime forest fires:** A. Park Williams, "Observed Impacts of Anthropogenic Climate Change on Wildfire in California," *Earth's Future*, July 15, 2019.

24 **In New Jersey:** "What Climate Change Means for New Jersey," US Environmental Protection Agency, 2016.

24 **Colorado ski towns:** Diana Olick, "Climate Change Is Taking a Toll on the $20 Billion Winter Sports Industry—and Swanky Ski Homes Could Lose Value," CNBC, July 5, 2019.

24 **heat-related deaths:** Daksha Slater, "Can Phoenix Remain Habitable?" *Sierra*, January 2, 2019.

24 **even in Arkansas:** "What Climate Change Means for

Arkansas," US Environmental Protection Agency, 2016.

24 **a wealth of investigative reporting:** See, for instance "Finalist: InsideClimate News," The Pulitzer Prizes, 2016, www.pulitzer.org/finalists/insideclimate-news; Sara Jerving, Katie Jennings, Masako Milissa Hirsch, and Susanne Rust, "What Exxon Knew About the Earth's Melting Arctic," *Los Angeles Times* and Columbia Journalism School's Energy and Environmental Reporting Project, October 9, 2015; Nathaniel Rich, "Losing Earth: The Decade We Almost Stopped Climate Change," *New York Times Magazine*, August 1, 2018.

25 **Humble Oil:** H. R. Brannon Jr. et al, "Radiocarbon Evidence on the Dilution of Atmospheric and Oceanic Carbon by Carbon from Fossil Fuels," *Eos*, 1957.

25 **a scientist at Shell:** The 1965 report cites research by M. King Hubbert, a researcher at Shell.

25 **"there seems to be no doubt":** "Sources, Abundance, and Fate of Gaseous Atmospheric Pollutants," Stanford Research Institute, 1968, accessed through ExxonKnew.org.

25 **Exxon began hiring scientists:** Neela Banerjee, Lisa Song, and David Hasemyer, "Exxon's Own Research Confirmed Fossil Fuels' Role in Global Warming Decades Ago," *Inside Climate News*, September 16, 2015.

25 **"Present thinking":** Neela Banerjee, Lisa Song, and David Hasemyer, "Exxon's Own Research Confirmed Fossil Fuels' Role in Global Warming Decades Ago."

25 **largest supertankers:** Neela Banerjee, Lisa Song, and David Hasemyer, "Exxon's Own Research Confirmed Fossil Fuels' Role in Global Warming Decades Ago."

25 **"CO2 and Climate Task Force":** Neela Banerjee, "Exxon's Oil Industry Peers Knew About Climate Dangers in the 1970s, Too," *Inside Climate News*, December 22, 2015.

25 **"overall goal":** Neela Banerjee, "Exxon's Oil Industry Peers Knew About Climate Dangers in the 1970s, Too."

26 **"Not to be distributed externally":** Neela Banerjee, Lisa Song, and David Hasemyer, "Exxon's Own Research Confirmed Fossil Fuels' Role in Global Warming Decades Ago."

26 **Exxon abruptly defunded:** John H. Cushman Jr., "Exxon Made Deep Cuts in Climate Research Budget in the 1980s," *Inside Climate News*, November 25, 2015.

26 **NASA scientist testified:** Philip Shabecoff, "Global Warming Has Begun, Expert Tells Senate," *New York Times*, June 24, 1988.

26 **"Emphasize the uncertainty":** Katie Jennings, Dino Grandoni, and Susanne Rust, "How Exxon Went from Leader to Skeptic on Climate Change Research," *Los Angeles Times*, October 23, 2015.

28 **lobbying partnership:** Andrew C. Revkin, "Industry Ignored Its Scientists on Climate," *New York Times,* April 23, 2009.

28 **they named the Global Climate Coalition:** Peter C. Frumhoff, Richard Heede, and Naomi Oreskes, "The Climate Responsibilities of Industrial Carbon Producers," *Climatic Change*, 2015.

28 **public attacks:** Naomi Oreskes and Erik M. Conway, *Merchants of Doubt*. For one example, see this memo that the GCC circulated to the media after the 1995 IPCC report, which had no basis in reality: www.documentcloud.org/documents/5976315-1996-GCC-Scientific-Cleansing.html, accessed July 26, 2021.

28 **millions on advertising:** Alison Mitchell, "G.O.P. Hopes Climate Fight Echoes Health Care Outcome," *New York Times,* December 13, 1997.

28 **an array of front groups:** "The Global Climate Coalition: Big Business Funds Climate Change Denial and Regulatory Delay," Climate Investigations Center, March 25, 2019.

28 **Exxon's board**: Sara Jerving, Katie Jennings, Masako Melissa Hirsch, and Susanne Rust, "What Exxon Knew About the Earth's Melting Arctic."

28 **US Senate voted 95–0:** Senate Res.98, 105th Congress, www.congress.gov/bill/105th-congress/senate-resolution/98.

28 **"Victory will be achieved":** Global Climate Science Communications, Action Plan, www.documentcloud.org/documents/1676446-global-climate-science-communications-plan-1998.html.

29 **something went sideways:** John H. Cushman Jr., "Industrial Group Plans to Battle Climate Treaty," *New York Times,* April 26, 1998.

30 **scores of ads:** Geoffrey Supran and Naomi Oreskes, "What Exxon Mobil Didn't Say About Climate Change," *New York Times,* August 22, 2017.

30 **talk in Beijing:** Neela Banerjee, Lisa Song, and David Hasemyer, "Exxon's Own Research Confirmed Fossil Fuels' Role in Global Warming Decades Ago."

30 **annual meeting:** Sara Jerving, Katie Jennings, Masako Milissa Hirsch, and Susanne Rust, "What Exxon Knew About the Earth's Melting Arctic."

30 **"aggressive agendas":** David Hasemyer and John H. Cushman Jr., "Exxon Sowed Doubt About Climate Science for Decades by Stressing Uncertainty," *Inside Climate News*, October 22, 2015.

30 **except its own ventures:** Amy Lieberman and Susanne Rust, "Big Oil Braced for Global Warming While It Fought Regulations," *Los Angeles Times* and Columbia Journalism School's Energy and Environmental Reporting Project, December 31, 2015.

30 **making new profits as Arctic ice melted:** Sara Jerving, Katie Jennings, Masako Milissa Hirsch, and Susanne Rust, "What Exxon Knew About the Earth's Melting Arctic."

31 **harsh letter to Exxon:** David Hasemyer and John H. Cushman Jr., "Exxon Sowed Doubt About Climate Science for Decades by Stressing Uncertainty."

31 **Exxon said it would stop funding:** David Adam, "Exxon to Cut Funding to Climate Change Denial Groups," *Guardian*, May 28, 2008.

31 **greatest predictor:** Cary Funk and Brian Kennedy, "The Politics of Climate," Pew Research Center, October 4, 2016.

31 **partisan divide over climate change:** "As Economic Concerns Recede, Environmental Protection Rises on the Public's Policy Agenda," Pew Research Center, February 13, 2020.

31 **cunningly accurate:** Warren Cornwall, "Even 50-Year-Old Climate Models Correctly Predicted Global Warming," *Science*, December 4, 2019.

32 **and still conclude:** Alan Buis, "Why Milankovitch (Orbital) Cycles Can't Explain Earth's Current Warming," NASA, February 27, 2020.

32 **caused by solar flares:** Kasha Patel, "Four Decades and Counting: New NASA Instrument Continues Measuring Solar Energy Input to Earth," NASA, November 28, 2017.

32 **historic temperature records:** Alan Buis, "The Raw Truth on Global Temperature Records," NASA, March 25, 2021.

CHAPTER TWO

33 **nine-year-old named Izerman:** Michelle Mizner and I made an interactive documentary about Izerman and two other incredible kids for *FRONTLINE* and The GroundTruth Project. Watch at apps.frontline.org/the-last-generation/.

38 **crop yields will drop by half:** "What Climate Change Means for Oklahoma," US Environmental Protection Agency, 2016.

39 **eighty years that the Marshallese would arrive:** Linda Ann Allen, "Enid 'Atoll': A Marshallese Migrant Community in the Midwestern United States," PhD dissertation, University of Iowa, 1997.

42–43 **a pair of surveys:** For the first survey, see Eric Plutzer et al, "Climate Confusion Among U.S. Teachers," *Science,* February 12, 2016. The second survey was conducted in 2019, and as of summer 2021, its results have not yet been published. However, Plutzer and the National Center for Science Education shared an analysis of the 2019 survey with me.

44 **just 5 percent take Earth science:** Per correspondence with Sean Smith of Horizon Research, Inc.

45 **"community resistance":** Eric R. Banilower et al, "The National Survey of Science and Mathematics Education, 2018," Horizon Research, Inc.

45 **the biggest predictor:** Eric Plutzer and A. Lee Hannah, "Teaching Climate Change in Middle Schools and High Schools: Investigating STEM Education's Deficit Model," *Climatic Change*, July 25, 2018.

45 **nine out of ten science teachers are white:** Eric R. Banilower et al, "The National Survey of Science and Mathematics Education, 2018," p. 8.

45 **eight out of ten teachers:** "Characteristics of Public School Teachers," National Center for Education Statistics, accessed May 2021.

45 **just half of public school students:** "Racial/Ethnic Enrollment in Public Schools," National Center for Education Statistics, accessed 2021.

45 **Older white people are more likely to deny the human causes:** Cary Funk, "Key Findings: How Americans' attitudes About Climate Change Differ by Generation, Party, and Other Factors," Pew Research Center, May 26, 2021.

46 **369 middle-school students in North Carolina:** K. T. Stevenson et al, "How Climate Change Beliefs Among U.S. Teachers Do and Do Not Translate to Students," *PLoS ONE*, 2016.

46 **115 Oklahoma science teachers:** Nicole M. Colston and Toni A. Ivey, "(Un)Doing the Next Generation Science Standards: Climate Change Education Actor-Networks in Oklahoma," *Journal of Educational Policy*, March 3, 2015.

47 **traveled to Alaska:** "STEM Experience Report—Melissa Lau," PolarTREC, n.d., www.polartrec

.com/resources/report/stem-experience-report-melissa-lau.

CHAPTER THREE

49 **Origins Resource Association:** The group still maintains an active website, and describes itself as "a nonprofit organization of scientists, educators, and citizens concerned about what we see as the brainwashing of our society into an unquestioning belief in evolution. Our mission is to furnish resources to help counter this trend," http://68.20.1.166/originsresource/about.htm.

50 **meeting agenda:** Barbara Forrest, "Combating Creationism in a Louisiana School System," *Textbook Letter*, July–August 1997.

50 **twenty-five-person:** Donald Wayne Aguillard, "An Analysis of Factors Influencing the Teaching of Biological Evolution in Louisiana Public Secondary Schools," PhD dissertation, Louisiana State University, 1998.

51 **voted 23 to 2:** Donald Wayne Aguillard, "An Analysis of Factors Influencing the Teaching of Biological Evolution in Louisiana Public Secondary Schools."

51 **promising a lawsuit:** Barbara Forrest, "Combating Creationism in a Louisiana School System."

51 **threat they saw in Darwinism:** David Masci, "Darwin in America: The Evolution Debate in the United States," Pew Research Center, February 6, 2019.

51 **"man dies as the brute dies":** Genevieve Forbes Herrick and John Origen Herrick, *The Life of William Jennings Bryan* (Stanton, 1925), p. 262.

52 **Tennessee's Butler Act:** Steven Chermak and Frankie Y. Bailey, *Crimes and Trials of the Century, Vol. I* (Greenwood Press, 2007), chapter 6. Most of the details I recount about the Scopes trial were gleaned from this excellent book.

52 **hanging out at a drugstore:** Doug Liner, "George Rappalyea," 2004, law2.umkc.edu/faculty/projects/ftrials/scopes/SCO_RAPP.htm.

53 **Susan Epperson:** Guy Lancaster, "Epperson v. Arkansas," *CALS Encyclopedia of Arkansas*, encyclopediaofarkansas.net/entries/epperson-v-arkansas-2528/.

53 **the concept of "creation science":** Roger Lewin, "Where Is the Science in Creation Science?" *Science*, January 8, 1982, pp. 142–144.

53 **Dallas, Chicago, and Atlanta:** Sherri Schaeffer, "*Edwards v. Aguillard*: Creation Science and Evolution—The Fall of Balanced Treatment Acts in the Public Schools," *San Diego Law Review* 25, no. 829 (1988).

53 **lawmakers in dozens of states:** E. J. Larson, *Trial and Error: The American Controversy Over Creation and Evolution; 3rd Edition* (Oxford University Press, 2003).

53 **National Center for Science Education:** "Our History," National Center for Science Education, ncse .ngo/our-history.

53 **shaped their flock:** "Religious Right," The Association of Religion Data Archives, www.thearda.com /timeline/movements/movement _17.asp.

54 **the court ruled:** Vivian Hopp Gordon, "*Edwards v. Aguillard*," *Britannica*, accessed June 14, 2021, www.britannica.com/topic /Edwards-v-Aguilard.

54 **"wholesale attack by intellectuals":** "The 'Wedge Document': 'So What?'" Discovery Institute, https://www.discovery .org/m/2019/04/Wedge-Document -So-What.pdf, accessed August 2021.

54 **she and scientist Paul R. Gross:** Barbara Forrest and Paul R. Gross, *Creationism's Trojan Horse: The Wedge of Intelligent Design* (Oxford University Press, 2004).

55 **Eleven parents sued:** Laurie Goodstein, "Judge Bars 'Intelligent Design' from PA Classes," *New York Times*, December 20, 2005.

55 ***"cdesign proponentsists"*:** "Cdesign Proponentsists," National Center for Science Education, September 25, 2008, ncse.ngo /cdesign-proponentsists.

55 **The judge ruled:** Laurie Goodstein, "Judge Bars 'Intelligent Design' from PA Classes."

55 **US Marshals:** Dylan Segelbaum, "Dover ID Trial: Q&A with Judge John E. Jones III," *York Daily Record*, November 6, 2015,

55 **"We have entered a new front":** *Discovery Institute Views*, Winter 2006, www.discovery.org/m /2019/08/VIEWS-Winter-06.pdf.

55 **the Academic Freedom Act:** Nicholas J. Matzke, "The Evolution of Antievolution Policies After *Kitzmiller v. Dover*," *Science*, December 17, 2015. Most of the discussion in this chapter of Academic Freedom bills and how they evolved came from this elegant analysis.

56 **Ouachita Parish:** "Local Louisiana School Board Praised for Adopting Policy Protecting Teachers Who Teach Evolution Objectively," Discovery Institute, December 1, 2006.

57 **their school board adopted:** "Freedom to Teach," *Central City News*, September 13, 2012.

57 **expand the Ouachita policy statewide**: Joshua Rosenau, "Louisiana Enacts a New Creationist Law," *Reports*

of the National Center for Science Education, July-August 2008.

57 **legislative efforts:** Katie Worth, "A New Wave of Bills Takes Aim at Science in the Classroom," *FRONTLINE*, May 8, 2017.

58 **least religious contingent:** Ross Tilchin, "The Libertarian Challenge Within the GOP," Brookings, December 27, 2013.

58 **a "creation care" movement:** See, for instance, www.episcopalchurch.org/ministries/creation-care/ (Episcopalian), www.umc.org/en/what-we-believe/umc-topics/social-issues/creation-care (Methodist), lutheransrestoringcreation.org (Lutheran), creationcare.org/what-we-do/an-evangelical-declaration-on-the-care-of-creation.html (Evangelical), www.catholiccreationcare.com (Catholic).

58 **A 2016 study:** E. H. Ecklund et al, "Examining Links Between Religion, Evolution Views, and Climate Change Skepticism," *Environment and Behavior*, October 26, 2016.

59 **several environmentalist groups:** Lydia Bean and Steve Teles, "Spreading the Gospel of Climate Change: An Evangelical Battleground," *New America's Strange Bedfellows Series*, November 2015.

59 **Evangelical Climate Initiative:** Adelle M. Banks, "Some Evangelicals Launch Campaign Against Global Warming, Others Protest," Religion News Service, February 9, 2006.

59 **walking on water:** "The Future is Green," *Vanity Fair*, May 2006, archive.vanityfair.com/article/2006/5/the-future-is-green.

59 **"anchor group":** Bean and Teles, "Spreading the Gospel of Climate Change."

60 **refrain from taking a position:** "A Letter to the National Association of Evangelicals on the Issue of Global Warming," signed by more than 20 religious leaders under the letterhead of the Interfaith Stewardship Alliance. The ISA was renamed the Cornwall Alliance in 2007.

60 **asking for the resignation of Richard Cizik:** Laurie Goodstein, "Evangelical's Focus on Climate Draws Fire of Christian Right," *New York Times*, March 3, 2007.

61 **"heavily tilted toward skepticism":** Robin Globus Veldman, *The Gospel of Climate Skepticism: Why Evangelical Christians Oppose Action on Climate Change* (University of California Press, 2019), p. 164.

62 **pundits have been affiliated:** Robin Globus Veldman, *The*

Gospel of Climate Skepticism: Why Evangelical Christians Oppose Action on Climate Change , p. 169.

62 **testified before both houses:** See his bio at the Cornwall Alliance at cornwallalliance.org /about/who-we-are/.

62 **Glenn Beck's show:** "Glenn Beck: Dangers of Environmental Extremism," Fox News, October 18, 2010.

62 **the Discovery Institute:** The talk is mentioned on his Heartland Institute bio: www.heartland.org /about-us/who-we-are/e-calvin -beisner.

63 **uncovered several forays:** www.desmog.com/discovery -institute/.

64 **in twenty states:** According to my count, these bills have been proposed in Alabama, Arizona, Colorado, Florida, Indiana, Iowa, Kentucky, Louisiana, Maryland, Michigan, Missouri, Montana, North Dakota, New Mexico, Oklahoma, South Carolina, South Dakota, Tennessee, Texas, and Virginia. In many, they have been proposed several times.

64 **survey of 1,427 teachers:** Eric Plutzer, Glenn Branch, and Ann Reid, "Teaching Evolution in U.S. Public Schools: A Continuing Challenge," *Evolution: Education and Outreach,* June 9, 2020.

CHAPTER FOUR

66 **the 1980s:** "Developing Content Standards: Creating a Process for Change," CPRE Policy Briefs, October 1993, www2.ed.gov /pubs/CPRE/rb10stan.html.

66 **"something is seriously remiss":** "A Nation at Risk," National Commission on Excellence in Education, April 1983.

66 **high-stakes tests:** "No Child Left Behind: A Desktop Reference," US Department of Education, 2002, p. 9. www2.ed.gov/admins/lead /account/nclbreference/reference .pdf.

66–67 **"critically analyze aspects of evolutionary theory":** John G. West, "Ohio Allows Alternative," Discovery Institute, December 17, 2002.

67 **mentioned some aspect of the phenomenon:** Kim Kastens and Margaret Turrin, "What Are Children Being Taught in School About Anthropogenic Climate Change?"

68 **Then came swift adoptions:** Christopher Holland, "The Implementation of the Next Generation Science Standards and the Tumultuous Fight to Implement Climate Change Awareness in Science Curricula," *Brock Education Journal,* 2020.

68 **"political correctness":** Erik W. Robelen, "Science Standards

Draw Fire From Ed. Leader in Kentucky Senate," *Education Week*, May 24, 2013.

69 **an "F" grade:** "The State of State Science Standards, 2012," Fordham Institute.

70 **Rep. Mark McCullough asked:** Chris Mooney, "Audio: Why Oklahoma Lawmakers Don't Want Kids to Learn About Climate," *Mother Jones*, May 16, 2014.

70 **"global warming is the main concern":** John Timmer, "Time Runs Out on Move to Reject Oklahoma Science Education Standards," *ArsTechnica*, May 28, 2014.

70 **Wyoming's legislature:** ncse.ngo/ngss-unblocked-wyoming. Iowa: ncse.ngo/two-anti-ngss-bills-die-iowa. Louisiana: Will Sentell, "New Science Standards Clear Final Hurdle, Take Full Effect in 2018-19 School Year," *Advocate*, March 8, 2017. West Virginia: Ryan Quinn, "'Next Generation' or Not, Science Standards Coming to WV Schools," *Charleston Gazette-Mail*, May 29, 2016. New Mexico: ncse.ngo/victory-new-mexico-0.

71 **dogged fight took place in Idaho:** Katie Worth and John Hand, "Inside Idaho's Long Legislative Battle over Climate Change Education," *FRONTLINE*, December 20, 2019.

72 **"teach to the state standards":** Eric R. Banilower et al, "The National Survey of Science and Mathematics Education, 2018."

72 **spend significantly more time:** A. Lee Hannah and Danielle Christine Rhubart, "Teacher Perceptions of State Standards and Climate Change Pedagogy," *Climatic Change*, November 23, 2019.

75 **lawmakers in Connecticut floated a bill:** Glenn Branch, "A Novelty in Connecticut," National Center for Science Education, March 2, 2018.

75 **pro-climate-change instruction measures:** Arizona (coded red), California, Connecticut, Hawaii, Minnesota, New Hampshire, New Jersey, New York, Rhode Island, and Washington (all coded blue): Glenn Branch, "The Year in Pro-Climate-Change-Education Legislation," National Center for Science Education, April 15, 2020. Such legislation was also proposed in Michigan and Pennsylvania (both coded red) later that year.

76 **grade each state's science standards:** "Making the Grade? How State Public School Science Standards Address Climate Change," climategrades.org .

76 **colored based on whether the state's legislature:** We coded a state red or blue if both houses of their state legislature were won by Republican or Democratic

lawmakers, respectively, in three of the four elections between 2013 and 2019. Nebraska has a non-partisan state legislature, but we coded it as red because the state has voted for Republican gubernatorial and presidential candidates since NGSS was released. The District of Columbia has voted the opposite direction in its mayoral and presidential elections in that period, so we coded it blue.

76 **But they were not the norm:** Of the red-coded states, Wyoming, Alaska, North Dakota, Arkansas, Kansas, and Michigan received a B+ or better. Louisiana, Oklahoma, Tennessee, Arizona, Mississippi, Montana, Missouri, South Dakota, Wisconsin, North Carolina, Idaho, Utah, Indiana, Florida, Ohio, Alabama, Georgia, South Carolina, Pennsylvania, and Texas all received between a B and an F. Of the purple-coded states, Colorado, Iowa, Kentucky, Maine, and New Hampshire all received a B+ or better, while Minnesota received a B– and Virginia failed. All of the remaining states and the District of Columbia were coded blue, and they all received a B+ or higher.

77 **social studies standards:** The seven states that require instruction in social studies classes about recent climate change are Hawaii, Massachussetts, New Jersey, Arizona, Indiana, Oklahoma, and Mississippi. Those in which the subject is mentioned as an example are California, Maryland, Oregon, Rhode Island, Vermont, New York, Alabama, Kansas, Michigan, North Carolina, Utah, Nebraska, West Virginia, and New Hampshire. The remaining thirty states and DC do not mention it in their social studies standards. We fact-checked our results with those of the K12 Climate Action, an initiative by the Aspen Institute, which reviewed both science and social studies standards in 2020. That report can be found here: www.k12climateaction.org/img/K12-SPL20-StandardsOutline-Screen.pdf.

77 **survey of 832:** Rana Khalidi and John Ramsey, "A Comparison of California and Texas Secondary Science Teachers' Perceptions of Climate Change," *Environmental Education Research*, 2021.

77 **Six years after Oklahoman educators:** This section is based on the following reporting: Nuria Martinez-Keel, "Oklahoma School Standards Could Add Biological Evolution, expand on climate change," *Oklahoman*, February 29, 2020; audio of the hearing: sde.ok.gov/sites/default/files/board-meetings/2020/STBD%202-27-2020.mp3; Randy Krehbiel, "Oklahoma Legislature Met Only 36 Days This Session, Leaving Some Work Undone," *Tulsa World*, May 19, 2020; Aaron Brillbeck, "Oklahoma Legislature Wraps Up Strange Session, Overrides Number of Stitt's Vetoes," *News 9*, May 22,

2020; Brian Bobek's bio: sde.ok.gov/about/brian-bobek; Glenn Branch, "Evolution and Climate Change in Oklahoma's New Science Standards," National Center for Science Education, May 18, 2020.

CHAPTER FIVE

81 **a quarter of American middle-school science classrooms:** Eric R. Banilower et al, "The National Survey of Science and Mathematics Education, 2018," p. 137.

81 **At the start of an eight-page section devoted to the subject:** *Glencoe iScience: Course 2 Integrated, Student Edition* (McGraw Hill Education, 2017), pp. 668–675.

82 **A chapter titled "Earth's Atmosphere":** *Glencoe iScience, Course 2*, pp. 570–611.

82 **pros and cons of renewable energy:** *Glencoe iScience: Course 1 Integrated, Student Edition* (McGraw Hill Education, 2017), p. 156.

82 **One lesson on coral reefs:** *Glencoe iScience: Course 3 Integrated, Student Edition* (McGraw Hill Education, 2017), p. 701.

82 **another on polar bears:** *Glencoe iScience: Course 3*, p. 642.

82 **A lesson about Hurricane Katrina:** *Glencoe iScience: Course 2*, p. 621.

82 **"*is suspected of* contributing to global climate change":** *Glencoe iScience: Course 2*, p. 442.

82 **"*there is evidence* that present day Earth is undergoing a global-warming climate change":** *Glencoe iScience: Course 3*, p. 630.

82 **"More carbon dioxide in the atmosphere *might cause*":** *Glencoe iScience: Course 1*, p. 435.

82 **"Volcanic eruptions *affect* climate":** *Glencoe iScience: Course 3*, p. 552.

83 **claim never to use a textbook:** Eric R. Banilower et al, "The National Survey of Science and Mathematics Education, 2018," p. 137.

83 **Industry tracker Simba Information:** Per a January 2019 conversation with Simba Information senior analyst Kathy Mickey.

83 **79 percent of the textbooks:** Eric R. Banilower et al, "The National Survey of Science and Mathematics Education, 2018," p. 138.

83 **major middle-school science product:** Per email from McGraw Hill spokesman Tyler Reed.

84 **an urgent telegram:** This anecdote was described in Adam R. Shapiro, "Civic Biology and the Origin of the School Antievolution

Movement," *Journal of the History of Biology*, Fall 2008, p. 430.

84 **stopping "parasitic" families:** George William Hunter, *A Civic Biology: Presented in Problems* (American Book Company, 1914), p. 261.

84 **"five races or varieties of man":** George William Hunter, *A Civic Biology*, p. 196.

85 **Gruenberg's book described:** Benjamin Gruenberg, *Elementary Biology* (Ginn & Company, 1919), p. 490.

85 **"I'm a Christian mother":** Randy Moore, *Evolution in the Courtroom: A Reference Guide* (ABC-CLIO 2001), pp. 32 and 40.

85 **The commission demanded:** Patsy S. Ledbetter, "Crusade for the Faith: The Protestant Fundamentalist Movement in Texas," PhD dissertation, North Texas State University, 1975, pp. 164–166.

85 **willingly ejected Darwinian evolution:** Judith V. Grabiner and Peter D. Miller, "Effects of the Scopes Trial," *Science*, September 6, 1974.

85 **Just fifteen of the fifty-three:** Gerald Skoog, "Does Creationism Belong in the Biology Curriculum?" *American Biology Teacher*, January 1978.

85 **tiptoed around the field's organizing principle:** There are some important exceptions to this rule. In his paper "Ella Thea Smith and the Lost History of American High School Biology Textbooks," published in the Fall 2008 issue of *Journal of the History of Biology*, Ronald P. Ladouceur writes that Smith's popular biology book *Exploring Biology* taught evolution in a highly sophisticated way for the era, though it did so with very few uses of the word "evolution." He also posits that biology's racist and classist ideals of the era played an even larger role in how textbooks presented evolution than pressure from Christian fundamentalists did.

86 **Sputnik:** Adam Laats and Harvey Siegel, *Teaching Evolution in a Creation Nation* (University of Chicago Press, 2016), p. 36.

86 **1,400 letters, calls, and petitions:** Paul Anthony, "'Drenched with Evolution': Reuel Lemmons and Churches of Christ in the Texas Textbook Controversy of 1964," *Restoration Quarterly*, 2017.

86 **broadcast on national television:** Adam Laats and Harvey Siegel, *Teaching Evolution in a Creation Nation*, p. 41.

86 **racist disputes:** Joseph Moreau, *Schoolbook Nation: Conflicts over American History Textbooks from the Civil War to the Present* (University of Michigan Press, 2003), pp. 52–92.

87 **became infuriated:** Wade Goodwyn, "Textbook Watchdog Norma Gabler Dies," NPR, August 1, 2007.

87 **"children will be taught atheism":** Paul Anthony, "'Drenched with Evolution': Reuel Lemmons and Churches of Christ in the Texas Textbook Controversy of 1964."

87 **kept lobbying:** Paul Anthony, "'Drenched with Evolution': Reuel Lemmons and Churches of Christ in the Texas Textbook Controversy of 1964."

87 **words devoted to evolution:** Gerald Skoog, "The Coverage of Human Evolution in High School Biology Textbooks in the 20th Century and in Current State Science Standards," *Science & Education*, 2005.

88 **though *iScience* was first published in 2012:** This paragraph was checked with McGraw Hill spokesman Tyler Reed.

88 **the section's first paragraph:** *Glencoe iScience: Course 2*, p. 669.

89 **a sizable section on aerosols:** *Glencoe iScience: Course 2*, p. 671.

89 **one chapter he had reviewed included three paragraphs:** *Glencoe iScience: Course 2*, p. 474.

90 **equal space to natural and human causes:** *Glencoe iScience: Course 1*, p. 170.

90 **"Scientists *hypothesize* that human activities":** *Glencoe iScience: Earth & Space, Student Edition* (McGraw Hill Education, 2012), p. 612.

90 **middle-school science textbooks for Texas:** *ScienceFusion Texas, Grade 6* (Houghton Mifflin Harcourt, 2015).

91 **gen-ed version devotes two pages:** Edward J. Tarbuck and Frederick K. Lutgens, *Pearson Earth Science* (Pearson, 2017), pp. 602–603.

91 **advanced version devotes ten pages:** Edward J. Tarbuck and Frederick K. Lutgens, *Earth Science* (Pearson, 2018), pp. 621-630.

91 **"What is causing it?":** Edward J. Tarbuck and Frederick K. Lutgens, *Pearson Earth Science*, p. 587.

91 **Only a handful of academics:** I found just four studies: Soyoung Choi et al, "Do Earth and Environmental Science Textbooks Promote Middle and High School Students' Conceptual Development About Climate Change?" *Bulletin of the American Meteorological Society*, July 1, 2010; Diego Román and K. C. Busch, "Textbooks of Doubt: Using Systemic Functional Analysis to Explore the Framing of Climate Change in Middle-School Science Textbooks," *Environmental Education Research*, 2015; Yu-Long Chao et al, "The Effects of Earth Science Textbook Contents on

High School Students' Knowledge of, Attitude Toward, and Behavior of Energy Saving and Carbon Reduction," *Science Education International*, 2017; and Casey R. Meehan, Brett L. M. Levy, and Lauren Collet-Gildard, "Global Climate Change in U.S. High School Curricula: Portrayals of the Causes, Consequences, and Potential Responses," *Science Education*, March 15, 2018.

91 **four California middle-school science textbooks:** Diego Román and K. C. Busch, "Textbooks of Doubt: Using Systemic Functional Analysis to Explore the Framing of Climate Change in Middle-School Science Textbooks."

91 **A 2018 study found such pussyfooting:** Casey R. Meehan, Brett L. M. Levy, and Lauren Collet-Gildard, "Global Climate Change in U.S. High School Curricula: Portrayals of the Causes, Consequences, and Potential Responses."

92 **"once part of a forest in Brazil":** *Glencoe iScience: Course 2*, p. 277.

92 **the rising populations of some countries:** *Glencoe iScience: Course 2*, p. 674.

92 **453 middle- and high-school students:** K. C. Busch, "Textbooks of Doubt, Tested: The Effect of Denialist Framing on Adolescents' Certainty About Climate Change," Environmental Education Research, 2021.

93 **Texas and California editions:** Dana Goldstein, "Two States. Eight Textbooks. Two American Stories," *New York Times*, January 12, 2020.

94 **coffin maker and history buff:** Harriet Tyson-Bernstein, *A Conspiracy of Good Intentions: America's Textbook Fiasco* (Council for Basic Education, 1988), pp. 7–8.

94 **One evening in November 2013:** Dan Quinn, "Oil and Gas Industry Advocates Launch Surprise Attack on Texas Environmental Science Textbook," Texas Freedom Network, November 21, 2013.

94 **the board voted to adopt:** Dan Quinn, "Big News: Science Education Advocates Thwart Late-Night Hijacking of Texas Science Textbook Adoption," Texas Freedom Network, November 22, 2013.

98 **no evidence of warming:** Jonathan D. Kahl et al, "Absence of Evidence for Greenhouse Warming over the Arctic Ocean in the Past 40 years," *Nature*, January 28, 1993.

101 **"our commitment to academic integrity":** Emailed statement from McGraw Hill spokesman Tyler Reed.

102 **shows up in a single lesson:** *Inspire Science: Grade 6 Integrated,*

Student Edition (McGraw Hill Education, 2020) Unit 4: Human Impact on the Environment, pp. 79–104.

102 **emphasizes deforestation:** *Inspire Science: Grade 8 Integrated, Student Edition* (McGraw Hill Education, 2020), Unit 4: Humans and Their Place in the Universe, p. 37.

102 **"due to natural and human activities":** *Inspire Science: Grade 7 Integrated, Student Edition* (McGraw Hill Education, 2020), Unit 4: Interactions Within Ecosystems, p. 55.

103 **deduce the relationship:** *Inspire Science: Grade 6 Integrated, Student Edition* (McGraw Hill Education, 2020), Unit 4, p. 90.

CHAPTER SIX

105 **"0.01 percent":** As I was fact-checking this book, I asked Miller for her sources for this statistic, and she said she didn't remember saying it and perhaps worded it wrong. "Even though we're the second largest producer of CO2 emissions, one country shutting it all down isn't going to do it, is the point I was trying to make," she said. "It's everybody, and especially China and India. If they don't join us, we can do all kinds of stuff and it's not gonna really move the bar."

107 **"well-directed program":** Jie Jenny Zou and Joe Wertz, "Oil's Pipeline to America's Schools," *StateImpact Oklahoma* and The Center for Public Integrity, June 15, 2017.

107 **children's "negative" attitudes:** Sheila Harty, *Hucksters in the Classroom: A Review of Industry Propaganda in Schools* (Center for Study of Responsive Law, 1979), p. 12.

107 **comic books about energy conservation:** Emily Atkin, "When Exxon Used Mickey Mouse to Promote Fossil Fuels," *HEATED*, March 5, 2020.

108 **the pair get in trouble:** "Mickey Mouse and Goofy Explore Energy," Exxon USA and Walt Disney Educational Media Company, 1976.

108 **included as an insert:** Sheila Harty, *Hucksters in the Classroom*, p. 48.

108 **a twenty-six-minute film:** *The Kingdom of Mocha*, accessed via YouTube: www.youtube.com/watch?v=PM1lxHXueik.

108 **more than 20 million schoolchildren:** "Windows of Opportunity: How Business Invests in U.S. Hispanic Markets, Vol. I," Hispanic Policy Development Project, Inc., 1987, p. 35.

109 **Massachusetts utility Eversource:** Steve Leblanc, "Utility-Backed Natural Gas

Booklets Spark Backlash at School," The Public's Radio, May 10, 2021.

109 **supported and sanctioned:** Jie Jenny Zou and Joe Wertz, "Oil's Pipeline to America's Schools."

111 **reached an estimated 3.3 million students:** 2019 OERB Update to the Oklahoma State Senate, appropriation.oksenate.gov/SubSelectAgencies/Agencies/2019/2019%20OERB%20Update.pdf.

111 **Ninety-eight percent of Oklahoma school districts:** Gaby Galvin, "Big Oil, Small Schools," *U.S. News & World Report,* August 25, 2017.

111 **17,000 teachers:** "Oklahoma Business Briefs for Jan. 31, 2020: OERB Hits Training Milestone," *Oklahoman,* January 31, 2020.

111 **a federal grant seeded a program:** Asher Price, "How a Natural Gas Group Pushed for New Energy Curriculum in Texas," *Austin American-Statesman,* July 10, 2018.

112 **"It is estimated that all human activity:** http://web.archive.org/web/20030819225830/http://www.classroom-energy.org/teachers/resources/climate.html.

112 **humans were responsible:** According to the monthly data that NOAA collects, in 2001, the atmosphere's carbon dioxide concentration was about 370 ppm; in 2021, it is about 420. The highest levels of carbon dioxide in human history prior to industrialization were about 300 ppm. Thus, in 2001, 70 ppm (19 percent) could be attributable to humans and in 2021, 120 ppm (28 percent) could be.

112 **The Rocky Mountain Coal Mining Institute:** www.rmcmi.org/education/global-warming-quiz, accessed through the Internet Archive's Wayback Machine.

113 **"What's more important":** www.ak.blm.gov/getenergized/start.html, accessed through the Internet Archive's Wayback Machine.

113 **booklet including six lesson plans:** Holly Lippke Fretwell and Brandon Scarborough, "Understanding Climate Change Lesson Plans for the Classroom," Fraser Institute, 2009, pp. 11 and 14.

114 **the US Department of Energy's site for teachers:** "Energy Literacy: Essential Principles for Energy Education," US Department of Energy, March 2017.

114 **Rocky Mountain Coal Mining Institute's educational materials:** www.rmcmi.org/education, accessed July 26, 2021.

114 **dozens of educational resources:** www.blm.gov/learn/interpretive-centers/campbell

-creek-science-center, accessed July 26, 2021.

114 **eight degrees Fahrenheit warmer:** The Alaska Climate Research Center, climate.gi.alaska .edu/ClimTrends/Change /TempChange.html.

114 **National Energy Education Development Project:** www.need .org.

115 **"Energy Infobooks":** The books can all be found on their site, and are sorted by grade and subject: www.need.org/need-students /energy-infobooks/, accessed July 2021.

116 **"Energy Stories and More":** https://web.archive.org /web/20150228015628/http:// issuu.com/theneedproject/docs /energystoriesandmore, first reported by Jie Jenny Zou and Joe Wertz in "Oil's Pipeline to America's Schools."

118 **the recipe book *Cooking with Dr. Pepper*:** Sheila Harty, *Hucksters in the Classroom*.

118–119 **only 700 acceptable for use in schools:** Michael Melia, "Many Online Climate Change Lessons Are Actually Junk," *Associated Press*, May 15, 2019.

119 **as many as two-thirds of American bird species:** "Survival by Degrees: 389 Bird Species on the Brink," National Audubon Society, October 10, 2019.

119 **"pristine nature":** In our fact-checking call, Miller told me that with this part of her presentation, she was trying to summarize the arguments of Alex Epstein, author of the book *The Moral Case for Fossil Fuels*. Epstein's argument has been embraced by the climate denial movement and criticized by environmentalists. The author was a keynote speaker at Heartland's 2018 "America First Energy Conference."

CHAPTER SEVEN

121 **increase in sea ice:** Axel J. Schweiger, Kevin R. Wood, and Jinlun Zhang, "Arctic Sea Ice Volume Variability over 1901–2010: A Model-Based Reconstruction," *Journal of Climate*, August 2, 2019.

122 **models have been remarkably accurate:** Zeke Hausfather et al, "Evaluating the Performance of Past Climate Model Projections," *Geophysical Research Letters*, December 2019.

122 **the notion that the Sun:** "Is the Sun Causing Global Warming?" Global Climate Change, NASA, climate.nasa.gov/faq/14/is-the -sun-causing-global-warming/.

122 **revealing that nearly half:** Eric Plutzer et al, "Climate Confusion Among U.S. teachers."

123 **mail a book:** Craig D. Idso, Robert M. Carter, and S. Fred Singer, *Why Scientists Disagree*

About Global Warming (The Heartland Institute, 2016).

123 **every science teacher in America:** Katie Worth, "Climate Change Skeptic Group Seeks to Influence 200,000 Teachers," *FRONTLINE,* March 28, 2017.

123 **education secretary Betsy DeVos:** Katie Worth, "DeVos Is Questioned About Campaign to Influence Climate Change Education," *FRONTLINE,* June 8, 2017.

123 **ranking House Democrats:** Katie Worth, "Democrats Condemn Climate Change Skeptics for Targeting Teachers," *FRONTLINE,* April 12, 2017.

123 **more than 200,000:** Katie Worth, "Dueling Books Compete to Educate Kids on Climate Change," *FRONTLINE,* December 2, 2018.

125 **systematically deflect attention:** Naomi Oreskes and Erik M. Conway, *Merchants of Doubt.*

125 **a small community of scientists:** Naomi Oreskes and Erik M. Conway, *Merchants of Doubt,* p. 6.

125 **solid-state physicist Frederick Seitz:** Dennis Hevesi, "Frederick Seitz, Physicist Who Led Skeptics of Global Warming, Dies at 96," *New York Times,* March 6, 2008.

125 **helped them distribute $45 million:** Naomi Oreskes and Erik M. Conway, *Merchants of Doubt,* p. 29.

125 **S. Fred Singer:** John Schwartz, "S. Fred Singer, a Leading Climate Change Contrarian, Dies at 95," *New York Times,* April 11, 2020.

126 **investigating the problem of acid rain:** Naomi Oreskes and Erik M. Conway, *Merchants of Doubt,* pp. 85–89.

126 **"nuclear winter theory":** Naomi Oreskes and Erik M. Conway, *Merchants of Doubt,* p. 63.

126 **questioned the science linking asbestos:** Amy Ridenour, "Lies, Damned Lies and Science: When Scientific Research Is Actually Scientific Sham," *National Policy Analysis,* February 2002.

126 **accused those scientists of self-interest:** Naomi Oreskes and Erik M. Conway, *Merchants of Doubt,* p. 128.

126 **stopped conducting original research:** Naomi Oreskes and Erik M. Conway, *Merchants of Doubt,* p. 270.

127 **the American Petroleum Institute hosted a meeting:** In response to questions about the 1998 meeting and other efforts to promote climate denialism, an API spokesperson responded: "The world looks a lot different today than it did twenty to thirty years ago when the internet and smartphones weren't even widely

available. Innovation has not only fundamentally changed life as we know it, it has transformed American energy to be produced cleaner than ever before and helped reduce CO_2 emissions to their lowest levels in a generation. Just like climate science has evolved considerably in recent decades, our industry has continuously evolved and is tackling the climate challenge head-on by investing in innovation and cleaner fuels, and supporting market-based government policies designed to build on this progress. The right response to climate change is to act based on the best science we have available to us today, and while others may stay focused on the past, we are focused on the future by taking action on climate while delivering the energy the world needs."

127 **eight-page memo:** The 1998 memo was uploaded by *DeSmog* to Document Cloud and can be seen here: www.documentcloud.org/documents/1676446-global-climate-science-communications-plan-1998.html.

128 **"junk science":** Sheldon Rampton and John Stauber, "How Big Tobacco Helped Create 'the Junkman,'" *PR Watch*, Third Quarter 2000; Naomi Oreskes and Erik M. Conway, *Merchants of Doubt*, pp. 150–151.

128 **connections with the tobacco industry**: David Helvarg, "Meet the Anti-Science Extremist Who Could Transform the EPA," *Nation*, November 18, 2016.

128 **industry lobbying group Global Climate Coalition:** Global Climate Coalition: Climate Denial Legacy Follows Corporations," Climate Investigations Center, April 25, 2019, climateinvestigations.org/global-climate-coalition-industry-climate-denial/.

130 **were in fact executed:** Graham Readfearn, "What Happened to the Lobbyists Who Tried to Reshape the US View of Climate Change?" *Guardian*, February 27, 2015; and Ben Jervey, "Fossil Fuel Industry's Global Climate Science Communications Plan in Action: Polluting the Classroom," *DeSmog*, February 27, 2015.

130 **An enterprising researcher:** Suzanne Goldenberg, "Work of Prominent Climate Change Denier Was Funded by Energy Industry," *Guardian*, February 21, 2015.

131 **In at least eight studies:** Justin Gillis and John Schwartz, "Deeper Ties to Corporate Cash for Doubtful Climate Researcher," *New York Times*, February 21, 2015.

131 **$10,000 plus expenses:** Ian Sample, "Scientists Offered Cash to Dispute Climate Study," *Guardian*, February 2, 2007.

131 **94 percent of stories:** Maxwell T. Boykoff and Jules M.

Boykoff, "Balance as Bias: Global Warming and the US Prestige Press," *Global Environmental Change*, 2004.

131 **1,768 press releases:** Rachel Wetts, "In Climate News, Statements from Large Businesses and Opponents of Climate Action Receive Heightened Visibility," *PNAS*, August 11, 2020.

133 **NSTA launched a website:** http://web.archive.org/web/20030114143220fw_/http://www.nsta.org/energy/index.html.

134 ***The Search for Solutions*:** When asked about the films and the teacher's guide, a ConocoPhillips spokesman said that ConocoPhillips had long provided charitable support for the creation and distribution of educational materials from the NSTA and other organizations, a practice they stopped in 2012 "due to the evolution of technologies providing digital content delivery." They noted that the NSTA has said that ConocoPhillips had no input on the materials, and that since 2003, the company has acknowledged the reality of anthropogenic climate change.

134 **teacher's guide:** Sean Cavanaugh, "2003 Teacher's Guide Prompts New Criticism of NSTA," *Education Week*, December 12, 2006.

134 **a blunt report:** Committee on the Science of Climate Change, "Climate Change Science: An Analysis of Some Key Questions," National Academies Press, 2001.

135 **The book was a collaboration:** The link between the NSTA and the ELC was first reported by Ben Jervey, "Fossil Fuel Industry's Global Climate Science Communications Plan in Action."

135 **"additional CO2 will be a net benefit":** co2coalition.org/frequently-asked-questions/.

135 **module on global climate change:** Environmental Literacy Council & National Science Teachers Association, "Resources for Environmental Literacy: Five Teaching Modules for Middle and High School Teachers" (NSTA, 2007).

135 **in a *Washington Post* editorial:** Laurie David, "Science a la Joe Camel," *Washington Post*, November 26, 2006.

136 **The editorial scandalized:** Sean Cavanagh, "Controversy Surrounds Science Group's Stance Against Distribution of Gore Film," *Education Week*, November 30, 2006.

136 **released a flurry of press releases:** Wheeler's press releases are archived at http://web.archive

.org/web/20101109100232/http://www3.nsta.org/pressroom.

137 **NSTA leadership issued a statement:** www.nsta.org/nstas-official-positions/teaching-climate-science.

138 **it disclosed that conclusion to shareholders:** Representatives of ExxonMobil did not respond to questions about the company's role in promoting climate denialism from the 1980s to the 2000s.

138 **Even the American Petroleum Institute:** Amy Harder, "Big Oil Lobby Showing Subtle Shifts on Climate Change," Axios, December 11, 2019.

138 **received funding from the tobacco industry:** Jessica Glenza, Sharon Kelly, and Juweek Adolphe, "Free-Market Groups and the Tobacco Industry—Full Database," *Guardian,* January 23, 2019.

138 **asserts that the health risks:** see www.heartland.org/Alcohol-Tobacco/Smokers-Lounge/index.html.

139 **received at least $650,000:** Neela Banerjee, "How Big Oil Lost Control of Its Climate Misinformation Machine," *Inside Climate News*, December 22, 2017.

139 **backlash over a billboard:** Rachel Nuwer, "Heartland Pulls Billboard on Global Warming," *New York Times,* May 4, 2012.

139 **confidential internal Heartland documents:** Brendan Demelle and Richard Littlemore, "Evaluation Shows 'Faked' Heartland Climate Strategy Memo Is Authentic," *DeSmog*, February 22, 2012.

140 **external review:** Suzanne Goldenberg, "Peter Gleick Reinstated by Pacific Institute Following Heartland Exposé," *Guardian*, June 7, 2012.

140 **11,250 schools:** "Global Warming Skeptics Target Students," UPI, May 5, 2008.

140 **presidents of 14,000 public school boards:** Sara Reardon, "Climate Change Sparks Battles in Classroom," *Science*, August 5, 2011.

141 **tallied the budgets:** Robert J. Brulle, "Institutionalizing Delay: Foundation Funding and the Creation of U.S. Climate Change Counter-Movement Organizations," *Climatic Change,* January 25 2013.

141 **$2 billion was devoted to persuading:** Robert J. Brulle, "The Climate Lobby: A Sectoral Analysis of Lobbying Spending on Climate Change in the USA, 2000 to 2016," *Climatic Change*, July 19, 2018.

142 **30 percent of Americans:** Anthony Leiserowitz et al, "Climate

Change in the American Mind: November 2019," Yale Program on Climate Change Communication, December 17, 2019.

EPILOGUE

144 **a subpar job:** Eric Plutzer and A. Lee Hannah, "Teaching Climate Change in Middle Schools and High Schools: Investigating STEM Education's Deficit Model."

145 **one-stop portal:** Find it at www.cleanet.org.

145 **The publishers of a book:** Ingrid H. H. Zabel, *The Teacher-Friendly Guide to Climate Change* (Paleontological Research Institution, 2017). Information about their mailing campaign can be found at www.givegab.com/nonprofits/paleontological-research-institution-and-its-museum-of-the-earth-and-cayuga-nature-center/campaigns/climate-change-challenge.

145 **high-school assemblies**: "Our Climate Our Future Live," Alliance for Climate Education, acespace.org/assembly/.

145 **2021 Pew survey:** Cary Funk, "Key Findings: How Americans' Attitudes About Climate Change Differ by Generation, Party and Other Factors."

146 **forty-nine studies on climate curricula:** Martha C. Monroe et al, "Identifying Effective Climate Change Education Strategies: A Systematic Review of the Research," *Environmental Education Research*, January 2017.

148 **most trusted institutions:** Nicole M. Krause et al, "Trends—Americans' Trust in Science and Scientists," *Public Opinion Quarterly*, September 24, 2019.

148 **Dig into those numbers:** Cary Funk, Brian Kennedy, and Courtney Johnson, "Trust in Medical Scientists Has Grown in U.S., but Mainly Among Democrats," Pew Research Center, May 21, 2020.

149 **scorned expert advice on masks:** Charlie B. Fischer et al, "Mask Adherence and Rate of COVID-19 Across the United States," *PLoS ONE*, 2021.

149 **then on vaccination:** Dan Keating et al, "Coronavirus Infections Dropping Where People Are Vaccinated, Rising Where They Are Not, *Post* analysis finds," *Washington Post*, June 14, 2021.

149 **childhood vaccines:** Dan M. Kahan, "Vaccine Risk Perceptions and Ad Hoc Risk Communication: An Empirical Assessment," Yale Law & Economics Research Paper # 491, January 27, 2014.

149 **genetically modified crops:** "The New Food Fights: U.S. Public Divides Over Food Science," Pew Research Center, December 1, 2016.

Columbia Global Reports is a publishing imprint from Columbia University that commissions authors to do original on-site reporting around the globe on a wide range of issues. The resulting novella-length books offer new ways to look at and understand the world that can be read in a few hours. Most readers are curious and busy. Our books are for them.

Carte Blanche:
The Erosion of Medical Consent
Harriet A. Washington

The Agenda:
How a Republican Supreme Court Is Reshaping America
Ian Millhiser

Reading Our Minds:
The Rise of Big Data Psychiatry
Daniel Barron

Freedomville: The Story of a 21st-Century Slave Revolt
Laura T. Murphy

In the Camps:
China's High-Tech Penal Colony
Darren Byler

The Subplot: What China Is Reading and Why It Matters
Megan Walsh